THE BEAUTY OF PHYSICS:
PATTERNS, PRINCIPLES, AND PERSPECTIVES

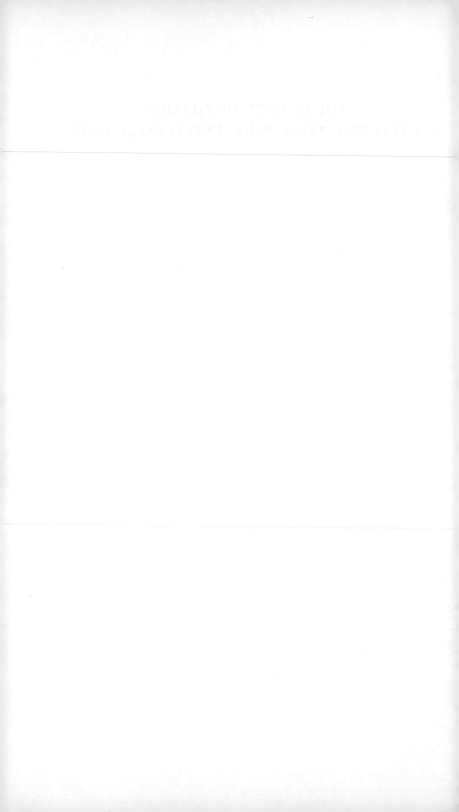

THE BEAUTY OF PHYSICS: PATTERNS, PRINCIPLES, AND PERSPECTIVES

A. R. P. Rau

*Department of Physics & Astronomy,
Louisiana State University, Baton Rouge*

OXFORD
UNIVERSITY PRESS

OXFORD
UNIVERSITY PRESS

Great Clarendon Street, Oxford, OX2 6DP,
United Kingdom

Oxford University Press is a department of the University of Oxford.
It furthers the University's objective of excellence in research, scholarship,
and education by publishing worldwide. Oxford is a registered trade mark of
Oxford University Press in the UK and in certain other countries

First Edition published in 2014

Impression: 1

Published in the United States of America by Oxford University Press
198 Madison Avenue, New York, NY 10016, United States of America

British Library Cataloguing in Publication Data

Data available

Library of Congress Control Number: 2014932181

ISBN 978–0–19–870991–6

Printed in Great Britain by
Clays Ltd, St Ives plc

Preface

The beauty of physics lies in its coherence in terms of a few fundamental concepts and principles. Even physicists have occasion to marvel at the overarching reach of basic principles and their ability to account for features stretching from the microscopic sub-atomic world to the cosmological expanses of the Universe. A few fundamental laws, principles, and ideas run throughout the subject. Even without their full mathematical implementation and detailed study, an initial understanding of the basic features of many phenomena can be grasped through qualitative applications of these fundamental principles.

Among these principles are not only the familiar laws such as those of motion, Newton's for classical physics or Schrödinger's for quantum physics, or laws of conservation, whether of energy, momentum or charge, that hold rigorously throughout physics, but 'themes' and 'metaphors' that arise many times in many sub-areas of the field.

The lay view of science as rigorous (as against speculative), proceeding step by step from observations and experiment to a theory in a strict, systematic way is a caricature, as every working scientist knows. The enormous roles of intuition, speculation, and guesswork in how we proceed, whether an ordinary scientist or the extraordinary Newtons and Einsteins, are not emphasized enough.

Even in mathematics, conjectures and working hypotheses are central, even if, in the spirit of rigour of the subject, the final product or theorem may be put in axiomatic form, every step connected to an already sound and established previous step or result. In science, where we are trying to understand the world around us, it is all the more important to recognize the role that intuition plays. In forming that intuition (the world around us plays a big role in this!) and how we use it in turn to generate more knowledge and understanding, we have our own ways of analogical thinking and metaphors.

Just as ordinary conversation is 'peppered' with metaphors, so too is physics. The dictionary defines metaphor as the use of a word or phrase to denote one kind of object for another by way of expressing an analogy between them. I use metaphor to mean equivalently analogy or, sometimes, a principle. The philosopher Jan Zwicky says that 'those who think metaphorically are enabled to think truly because the shape of their thinking echoes the shape of the world'. The great 19th-century physicist James Clerk Maxwell, who unified electricity and magnetism and showed how light/optics is also part of this union, contrasted metaphor or analogy with puns, another part of speech that we are familiar with and often, if groaningly, love: 'In a pun, two truths lie hid under one expression. In an analogy, one truth is discovered under two expressions'.

It is the discovery of one truth under two, or many, different realizations that underlies this book. One of the attractive features of physics is that the same simple principle applies across many phenomena that are seemingly very different. And a characteristic of physics is to see the world from different points of view, quantum physics in particular emphasizing 'complementary' representations leading to the same result.

Charles Darwin, another great scientist and a contemporary of Maxwell, also used 'selection' as a metaphor, that one cannot do science without being metaphorical, and said: 'No one objects to chemists speaking of "elective affinity", and certainly an acid has no more choice in combining with a base, than the conditions of life have in determining whether or not a new form be selected or preserved' [1]. Since science, with physics my main focus, tries to connect phenomena that at first sight appear widely different (falling apples and the Moon's orbit), by boiling them down to a small set of essential principles and laws, metaphor and analogy pervade our subject. Some, in particular, are so universal that we immediately jump to saying 'that is just a pendulum', completely akin to saying 'she is a rose', or 'Juliet is the Sun', in an everyday context.

Indeed, as familiar already in ordinary language as a powerful metaphor, consider the pendulum. Its swing from one extreme to the other is often invoked in social or economic contexts. Another simple example is the two-faced quality of a coin. But physics sees even further elements in them. Often, when we encounter a physical situation or a certain mathematical equation, we will see in it the pendulum even though there may be no actual pendulum, no strings or bobs. That identification invokes immediately all kinds of other implications and consequences, both in the mathematical analysis and for the physics of the subject under study.

In molecular vibrations, such as in the CO_2 molecule, the quantum motions of electrons and nuclei are metaphorically the pendulums. In electromagnetic radiation, including the visible light we observe, there are not even any concrete material particles, only electric and magnetic fields executing simple harmonic motion. But, to a physicist, they are all 'just a pendulum', adding further richness to these metaphors. It is some of this flavour that this book tries to convey.

Perhaps because of the way our brains are constructed and have evolved, there is often a tendency, in many religions or mythologies, or even in science, to seek a single theory of everything. Another tendency, whether in physics or today's biology, is to seek the fundamental at the lowest size, whether in genes and DNA/RNA in molecular biology or quarks and leptons in particle physics (what had previously been at the level of atoms or, next, of nuclei). Yet another is to see data, numbers, and statistics as the essence of science. While they are an important part, it is the fundamental concepts and principles, including the laws of motion and of conservation, that apply across the whole field, that define our subject. And, as emphasized by Newton, there are at every stage and every level of inquiry 'initial conditions', parameters, fundamental constants, and constructs that are also crucial for the relevant physics but to be taken as given.

While every physicist recognizes many such themes and principles, employing them as part of the very vocabulary of the subject, they are not often spelled out or brought together, especially for students during their courses of study. As with metaphors and themes more generally, students are expected to imbibe them with increased exposure and through their own encounters with them, sometimes applying and extending them in new contexts. In time, every physicist thereby develops a perspective on the subject that extends beyond the specific books and papers read or authored.

This book presents some of the principles and perspectives that dominate my view of the world of physics. The very use of the word 'perspective' signifies a subjective element and each physicist will have his or her own set of favourite topics. Inevitably, some of them will overlap and some not, reflecting one's own history in the subject and one's taste. This is as it should be, the subject itself larger than the sum of its parts. It should be no surprise that the themes I have chosen are ones that have played important roles in my own research career. Every chapter reflects this, the topics discussed having been of central interest to me in my research and teaching.

To whom do I address this book? Any student of physics in the senior undergraduate years, and certainly graduate students and researchers, will be able to follow the entire discussion. But I am aiming at a broader audience of the intellectually curious reader in other sciences and even outside the sciences and mathematics. In each chapter on a theme, later sections will deal with illustrations that will require exposure to some advanced physics for their full appreciation. But each chapter will present very simple examples in the beginning to illustrate the theme. These will be accessible to anyone willing to exercise his or her logic and imagination, and not shy away from thinking in terms of symbols and following simple algebraic relations between them.

Mathematical symbols, some simple algebra, and elements of calculus are inescapable, being an integral part of the language of physics. No one needs avoid them. Their usage in this book is deliberate and, in part, my contribution towards the debate initiated by C. P. Snow's *The Two Cultures and a Second Look* (1964). Educated and intellectually alive persons, whatever their expertise, have an appreciation of literature, music, and the arts. As one who appreciates music and the fine arts but without any training in them, I have had occasions, when in the company of someone versed in them, to realize how much more there is to understand and appreciate in a painting or a piece of classical music. Nevertheless, I already derive some benefit and satisfaction even when I do not have their aid.

It is in somewhat that spirit that I wish to communicate to my non-scientist reader some of the beauty and depth of the principles and patterns of my subject, even if some of their sophisticated realizations may be only for physicists. Just as in other subjects and fields of inquiry, physics and mathematics are also mostly about patterns and how they are put together, sometimes in unexpected but pleasing combinations and contexts. While most of the patterns and themes apply more broadly outside of physics, physics sees even further facets of them. The power and elegance of patterns and principles in exposing truths of the way things are, and why they are as they are, are what make physics part of the liberal arts as understood in the US college curriculum.

Equations are sparingly used in this book, especially at the beginning of chapters, and they are very simple and spelled out so that no prior acquaintance with them is necessary. A general reader is encouraged to continue to read on into the more sophisticated illustrations later in the chapter. Some advanced material is set off against a shaded background. Since the accent is on the theme and ideas, technical and mathematical details or equations are mostly avoided in the later sections of the chapter when discussing more sophisticated occurrences of the ideas. If a reader skips these sections, or gets only an impressionistic feel for the range and power of explanation with only physicists fully appreciating them, the theme itself will nevertheless have been developed for everyone through the first simple illustrative examples. At the same time, even physicists familiar with these later topics may appreciate their unification with others in a continuing theme from simple beginnings.

My motivation for writing this book is of course to convey some of the power, significance, and beauty of physics. Part of the inspiration for writing came from the powerful effect three books had on me, even on a first reading. While in no way in the same league as these authors and their books, this book and I have been influenced by their style and content. The book *The First Three Minutes*, by Steven Weinberg [2] was addressed in part to his 'intelligent lawyer friend', conveying to such a reader who is willing to follow a thread of logic and argument a sophisticated understanding of the very first minutes of the Universe as revealed by physics. Both that book, and Richard Feynman's *QED*, [3] while addressing a broad audience, make no concession in rendering the physics rigorously. While presenting accessibly and with little mathematics, there should be no compromise on the physics itself and portraying it accurately. The book by Rudolf Peierls, *Surprises in Theoretical Physics*, [4] is somewhat different, aimed at graduate students and researchers, and is full of the kind of insight and perspective on even well-known topics that students do not get normally in courses and from textbooks. I bow in acknowledgement to these three works.

A book such as this is not one where every item is footnoted and referenced, and notes and citations have been kept to a minimum. Brief biographical footnotes are given for every person mentioned. These are mainly for the reader unaware of these noteworthy people, to give them a minimum impression of who they were. Readers interested in delving more into those lives and works can, in this day and age, turn to Wikipedia, encyclopaedias, and further references therein. Regarding references in this book, all the physics mentioned will be familiar to research physicists. More junior students will also know of the standard textbooks in classical mechanics, electromagnetism or quantum mechanics to turn to if they want to learn more details. Since this is about my perspective, my own research work is naturally reflected throughout, and a few references to it are given where the interested reader can find further elaboration. Otherwise, references have been given only for a handful of very specifically mentioned items.

Acknowledgements

This book was written during a year of sabbatical leave from my home institution, Louisiana State University, Baton Rouge, and I thank it for its support. I also thank the following institutions and colleagues for their support and hospitality during this year: Alexander von Humboldt Stiftung, Germany; Technische Universität, Darmstadt, Germany (Dr Gernot Alber); Universidad de Santiago de Chile (Dr Juan Carlos Retamal); Universidad de Valparaiso, Chile (Dr Quintin Molina); Universidad de Concepcion, Chile (Dr Aldo Hidalgo); Universidad de Los Lagos, Osorno, Chile (Dr Alberto Gantz); Universidad de Magallanes, Punta Arenas, Chile (Dr Victor Diaz); Omora Field Station, Isla Navarino, Chile (Dr Jaime Jimenez); and Australian National University, Canberra, Australia (Drs Stephen Buckman and James Sullivan).

Over the years, I have bounced some of these ideas off my biologist wife, Dr Dominique Homberger, and we also shared the sabbatical trip. Among other places, historically iconic ones such as the Straits of Magellan, the Beagle Channel and Isla Navarino in Chile, and Captain Cook's Cape Tribulation on Australia's north-east coast were inevitably inspirational, especially in writing the chapter on maps.

My perspective on physics has been influenced by numerous classmates, teachers, and students that I have interacted with over five decades and I thank all of them. My special gratitude to my doctoral advisor, the late Dr Ugo Fano (University of Chicago), my post-doctoral mentor, the late Dr Larry Spruch (New York University), and to my colleagues and students in the Department of Physics and Astronomy at Louisiana State University. I learnt much from many discussions over decades with the late Dr Mitio Inokuti, who was similarly interested in ideas in physics. And discussions in recent years with Dr Gernot Alber (Darmstadt) of some of the topics discussed in the book have been very valuable.

My editor, Dr Sonke Adlung, has been patiently encouraging throughout since the very first germ of an idea of this book several years ago, and my grateful thanks to him. I thank also Ms Jessica White for the work in bringing this book to completion.

Contents

1

Adding a Dimension

1.1 Dimensions in Physics

The theme of this chapter, as indicated in its title, is adding a dimension. This is often useful in physics. But consider first what is meant by dimensions in physics. Physics is the study of the world around us in a disciplined way, with increasing precision and depth. But long before physics, even our earliest proto-human ancestors must have recognized the role that size, location, and distances between locations plays in the world and for their life in it. Length or distance along a line, a linear 'dimension', is therefore among the most primitive concepts for describing and understanding the world we live in. Also an early realization was that there are three different distances or displacements – forwards/backwards, sideways or left/right, and up/down – that we live in a three-dimensional world. There are three independent degrees of freedom in the motion of any object.

From this humble but important beginning, the concept of dimension has been extended in many ways, both in physics and in ordinary language, so as to have gained rich metaphorical uses in both. Thus, we talk of the many dimensions of an idea, construct or person, or of someone having an extra dimension to him or her, or of a well-rounded argument. Similarly in physics, all our quantities of interest are characterized in terms of basic dimensions that extend far beyond the original lengths, breadths, and heights we started with.

All separations and distances, no matter how small or large, from nanometres to kilometres or astronomically large distances, share in essence the common feature of being a linear dimension. They may all be thought of for that purpose as an [L], a length, this being the essential dimensional element that unifies and characterizes them. Units for measuring them may vary with context and country, from the initial hands and feet or stride lengths that gave natural, human measures for them, to the inches and cm (centimetres) and their multiples that different peoples instituted for precision dealings. In an equation in physics, no matter how complicated, the dimensional aspect of such elements is that the quantity is an [L].

The primitive beginnings of recognizing lengths, breadths, and heights must have stretched immediately to ideas of putting them together in multiplicative fashion, that areas are formed of two of these, and volumes by combing three in the three independent directions of the world we live in. Certainly with the advent of agriculture and settled civilization and the building construction that went with it, all peoples developed an understanding of areas and volumes, and the study of geometry (literally measurements on the Earth) predates physics as a subject. Mound and pyramid builders, Greeks, Arabs, and a myriad others developed sophisticated ideas in geometry, and instruments based on them, as part of their civilization and culture. A rich lexicon of terms – acres and hectares, gallons and litres – developed. But for physics, their essence lies in that two or three lengths in independent (mutually perpendicular or 'orthogonal') directions are put together multiplicatively so that all areas are dimensionally $[L]^2$ and all volumes $[L]^3$, using the mathematical notation of exponents for squares, cubes, etc. These dimensional aspects are shared by all areas or volumes, whatever other distinctions may apply to them and whoever measures them in their own distinct units. The volume of a cube of side a is just a^3 itself, while that of a sphere of that radius has additional, dimensionless factors (including the universal constant $\pi = 3.14159\ldots$) in its $(4\pi/3)a^3$ but both are equally of dimension $[L]^3$, a cubed length, such as m^3 or cm^3 (sometimes abbreviated as cc, a thousandth of a litre).

One can already see benefits from this kind of discipline and precision in thinking that characterize physics. Starting with little more than the above observation, and with [L] the only dimension to play with, along with its different powers, one can draw interesting conclusions about the world around us. Two similar wading birds, a flamingo and a stilt (Figure 1.1), have body masses, respectively, of about 2 kg and 120 g for a mass or, equivalently, volume ratio of, approximately, 16 between them. (There is here an implicit assumption, essentially correct, that all birds have about the same density so that masses are proportional to volumes.) Everything about the bigger flamingo will, of course, be comparatively larger than for the smaller stilt, but consider going further, to more precise terms. Based on how volumes relate to [L], we might expect that the lengths of their legs would be in the ratio $16^{1/3} \approx 2.5$,

Figure 1.1 Two birds, a stilt (left) and a flamingo (right), with similar wading habits and habitats. Text relates their sizes according to dimensional scaling. Tom Grey: <http://ic2.pbase.com/g6/44/316244/2/84224488.cb3EfssK.jpg>; William Duke: <http://www.pbase.com/photosbyduke/image/95530287>

in conformity with the observed 20 and 8 in, respectively, of their legs on average. Note that we took the 1/3 power (cube root) of 16, not 16 itself, in going from comparing volumes to lengths (of legs). With this one small logical step, we made quantitative sense of a little element of what we see around us, that these various numbers are not a mindless collection in some catalogue, but the ratio of masses bears a simple relationship to the ratio of leg lengths, reflecting the dimensionality of our world. Our example was deliberately from the biological world because whatever the enormous biological differences between flamingoes and stilts, and whatever the variations among individuals of either group, the animate world is just as constrained as the inanimate by the laws of physics, here of gravity.

That the surfaces and volumes of an object scale differently and, in particular, that the ratio surface/volume, which scales as $[L]^{-1}$, diminishes with increasing size permit non-trivial connections and understandings of the world around us, both physical and natural. (In scientific notation, negative powers indicate terms in the denominator.)

Note that both the surface and the volume of an object increase as the object is scaled up but the ratio decreases, as the volume increases faster. Doubling lengths increases areas by a factor of four, and volumes by a factor of eight, but the ratio of surface area to volume is halved. Indeed, Galileo[1], whom we might count as the first physicist, wrote eloquently about this, as did the later and, perhaps, the best expositor of this kind of 'dimensional analysis', D'Arcy Thompson[2]. Metabolism depends on the whole body volume but heat is lost or absorbed from the surface so that infants chill more rapidly than adults and, as every parent knows, need to be wrapped up more. The folded and broken up structure of lungs and intestines, organs that have to absorb through a surface, is Nature's solution to accommodating a large surface within the constraints of a given volume (Figure 1.2). Thus, in D'Arcy Thompson's eloquent words [5], 'the form of a body is a diagram of the forces acting upon it'.

Moving beyond the single, static length dimension [L] to considering objects in motion, that is, changes of their position in time, we have to introduce a new element, time. Time as an intrinsically new dimension, [T], enters as a natural extension of the concept of dimension. Today, in colloquial usage, even laymen associate the idea of time with the 'fourth dimension', linking it especially with Einstein[3] and the

[1] Galileo Galilei, 1564–1642, Italian. May be considered the first physicist for his careful observations of bodies in free fall and rolling down inclined planes, on the basis of which he arrived at the principles of inertia. In particular, he realized that the state of rest is not a special one, as had been thought previously, but that all uniform motion, including the case of zero velocity, continues in the absence of impressed forces. This was formalized later as Newton's First Law of Motion. The first telescopes were just appearing and Galileo developed them further, turning them to observing the heavens and discovering sunspots and the moons of Jupiter, both of philosophical import, in showing that heavenly bodies are not made of some unblemished quintessence different from on Earth and that there are other planetary systems analogous to our own Solar System. He also saw in swinging objects such as pendulums time-keeping devices, and recognized the role of dimensions and scaling to explain the world around us.

[2] D'Arcy Wentworth Thompson, 1860–1948, Scottish. Mathematician, biologist, and scholar of the classics. His book on the structure of plants and animals became a classic, also as a piece of literature. He emphasized the role of physical and mechanical laws in biology that was otherwise dominated by Darwinian selectionist thinking. The role of mathematics such as Fibonacci sequences and of geometrical transformations, and the way of thinking he introduced influenced many biologists and others who have followed him.

[3] Albert Einstein, 1879–1955, Swiss and American. Revolutionized physics and philosophy through several papers in 1905 on Special Relativity, Brownian motion (which demonstrates the underlying atomic structure of matter), and the photoelectric effect

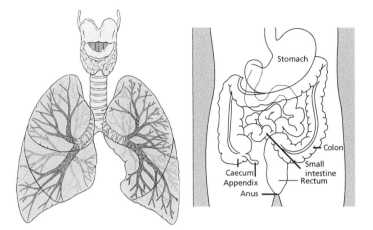

Figure 1.2 Lungs and intestines. Organs for absorbing oxygen or nutrients have folds and other broken-up structures to get a larger surface/volume ratio than they would have otherwise. Patrick J. Lynch: <http://en. wikipedia.org/wiki/File:Lungs_diagram_detailed.svg>. Indolences: <http:// en.wikipedia.org/wiki/File:Stomach_colon_rectum_diagram.svg>

Theory of Relativity. But, invoking it as another dimension, independent of space, goes all the way back to the beginnings of physics, again to Galileo. Indeed, the recognition of the pendulum clock as the basic apparatus for measuring time, and the pendulum equation for the time period of one full swing (see Figure 1.3), generally attributed to Galileo,

$$T = 2\pi \sqrt{\frac{\ell}{g}}, \tag{1.1}$$

led to the beginnings of our measuring and studying motion. The above pendulum equation is one of the first equations a student of physics learns in high school or college.

(a key step in the development of quantum physics), and a decade later with his General Theory of Relativity and Gravitation. His scepticism of the interpretation of quantum physics, and belief that it was incomplete, has drawn renewed attention in recent years in the field of quantum information, and his development of Bose–Einstein statistics and the prediction of a 'condensate' at very low temperatures has now been realized experimentally and is leading to many novel developments. An avowed pacifist, he spoke out (as in the Bertrand Russell–Einstein Manifesto) in his late years against nuclear war and weapons.

Figure 1.3 A simple pendulum. For small oscillations about the vertical equilibrium position, a string has a time period that depends only on its length, ℓ, and Earth's gravity, as per Eq. (1.1). It is independent of the shape and mass of the suspended bob.

The time period on the left-hand side of Eq. (1.1) depends only on the length, ℓ, of the pendulum divided by the acceleration due to gravity, g, a measure of how all masses fall under the attraction of Earth's gravity. A key idea of physics, one first recognized by Galileo and later incorporated into Einstein's General Theory of Relativity, is that the acceleration is independent of the mass of the falling body. The rate of change of position with time, which is a speed or velocity, has dimensions of $[L][T]^{-1}$ and the rate of change of velocity, an acceleration, involves one more time element in the denominator, to give it dimensions of $[L][T]^{-2}$. It is clear by inspection that Eq. (1.1) is dimensionally consistent, as indeed any equation must be. You cannot add apples and oranges, nor terms that do not match dimensionally. Indeed, the practical use of dimensional analysis as an aid to memory allows us to tell the student that, should there be confusion in recalling whether it is ℓ/g or g/ℓ inside the square root, thinking of dimensions will give the answer. The pure number 2π in the formula, being dimensionless, has of course to be memorized but the rest of Eq. (1.1) can be argued for strictly from dimensional considerations, without any further knowledge of physics. Identifying the length, and the constant g that characterizes the restoring force due to gravity that makes the pendulum swing, as the only relevant variables, the time period must perforce involve the combination $\sqrt{\ell/g}$ to give something of dimension $[T]$. No other combination will do.

Next, with just a few more steps of thought, we start making further sense of the world around us. When we walk, our legs swing in pendular fashion. Taking 1 m for that leg length (also, approximately, the length of our stride), and with $g \approx 10$ m/s^2, from Eq. (1.1) follows a typical walking speed of 1 m/s or 3.6 km/h (approximately 2 mph). That is a good estimate for our average walking speed. Dimensional analysis can also be used to argue that for objects moving through fluid media, whether birds and airplanes through air, or fish and submarines through water, their typical velocities, which are dimensionally length/time, must involve $\sqrt{\ell}$, that is, must scale with the square root of their lengths. Indeed, stalling speeds of a large airplane relative to that of a small one scale in this fashion, as do swimming speeds of fish.

Consider tsunamis, such as those in 2004 in the Indian Ocean and in 2011 in Japan. One of their awesome features is that they race across oceans at speeds often of the order of 150 m/s or 350 mph (the speed of a jet airplane). This can again be understood in simple terms, ignoring all the complicated hydrodynamics of flow (still one of the most difficult of topics in physics), through dimensional considerations alone. When a major disturbance on the sea floor displaces a huge volume of water, the ocean surface will return to equilibrium by that piled-up volume spreading out. The restoring agent is again gravity (in this context of large volumes; for water waves in shallow bowls or ripple tanks, it would be a different agent, surface tension) so that we expect g to be involved. The tsunami speed, being dimensionally $[L][T]^{-1}$, as is any speed, needs a length dimension, $[L]$, which upon multiplying with g, which is in dimensions of $[L][T]^{-2}$, and taking a square root will provide a candidate expression for it. The natural choice for a length involved is the ocean depth or the wavelength of the waves, both approximately a few km. With these inputs, and a 2 km depth taken as a characteristic average for our oceans, $\sqrt{2000 \times 10}$ is indeed, approximately, 150 m/s or 350 mph.

Moving on to the subject of mechanics, the study of the motion of physical objects, the next element is their mass, a new dimension. While irrelevant for falling under gravity (a statement of great import about the nature of gravity), other motions in mechanics do depend on the mass of the moving object. Indeed, mass is the measure of how much an object resists forces that try to change its state of motion. Being intrinsically different from length and time, we introduce $[M]$ for this, the amount of stuff in the object. Again, whether we measure in kg or tonnes, whether the microscopic mass of an electron or the mind-boggling mass of a black hole, all are dimensionally $[M]$. The three dimensions, [L], [T], and [M], together suffice to describe all physics as Galileo and Newton[4] knew it, all of mechanics then and since.

Only the extension into electromagnetism, which came in the mid-19th century, required supplementing [L], [T], and [M] with one more independent dimension, that of charge, which may be denoted [Q]. All other electrical quantities such as current or voltage as well as magnetic quantities can be expressed in terms of the dimension of charge combined with the above three dimensions of mechanics. That only one new dimension is needed is in part an expression of an important aspect of the unified nature of all electricity and magnetism. There is freedom in that choice of element, whether charge or current (of dimensions charge/time) or voltage, the essential aspect being that, along with the three for mechanics, a combination of four dimensions can describe all of mechanics and electromagnetism. And this is as true of today's quantum physics as it was in the classical physics of Newton and Maxwell[5].

Consideration of mass leads naturally to the concept of density of material bodies, mass/volume, with dimension $[M][L]^{-3}$. All quantities of mechanics — momentum, force, energy, pressure, etc. — are various combinations of the same three dimensions, respectively, $[M][L][T]^{-1}$, $[M][L][T]^{-2}$, $[M][L]^2[T]^{-2}$, and $[M][L]^{-1}[T]^{-2}$. Note in the second an expression of Newton's famous (second) law of motion, $F = ma$, that force is mass times acceleration or, alternatively, the rate of change

---

[4] Isaac Newton, 1642–1727, English. One of the greatest scientists of all time and the founder of physics through his discovery of the laws of motion and of gravitation, and of mathematical analysis as one of the developers of the infinitesimal calculus. He also developed the subject of geometrical optics for light propagation, including the design of telescopes. His concept of time as an absolute background flow against which to view all phenomena, despite later modifications in Einstein's Theory of Relativity, continues to be problematical to laymen, physicists, and philosophers (see Chapter 7). His laws of motion, despite the later revolutions of relativity and quantum physics, continue to be relevant, from our mundane bicycles and automobiles to sophisticated space missions.

[5] J. C. Maxwell, 1831–1879, Scottish. Formulated a unified theory of electric and magnetic phenomena through the basic set of equations of classical electromagnetism. In doing so, he concluded that waves of electric and magnetic fields, propagating at the speed of light, c, must exist. Visible light and optics are a part of this electromagnetic 'spectrum' that also embraces other waves, such as ultraviolet, x-ray, gamma ray, microwave, and radio waves. This formulation by him and his followers was crucial for Einstein in his quest to make mechanics and electromagnetism compatible, thereby leading to the Special Theory of Relativity, with c a universal constant and a fundamental re-interpretation of space and time. Maxwell also developed the kinetic theory of gases, another fundamental piece of physics, one relating microscopic statistical mechanics to macroscopic thermodynamics.

of momentum (which is mass times velocity or $[M][L][T]^{-1}$) in time, $F = dp/dt$ as rendered in Newton's differential calculus. The kinetic energy of motion, $mv^2/2$, or the gravitational potential energy, mgh, of a mass at a height h above the surface of the Earth, or any other form of energy, including the famous Einsteinian relation, $E = mc^2$, with c the speed of light, are always a $[M][L]^2[T]^{-2}$. Again, in any equation of physics, such as Newton's law of motion or Einstein's equivalence of energy and mass, the left- and right-hand sides and all terms in them must be dimensionally consistent.

As one more example of the power of dimensional analysis to give non-trivial results, consider a problem unfamiliar to most, the question of how the fireball from a large explosion, whether chemical or nuclear, expands with time. When one views such an explosion on a TV screen, it would seem the domain of a specialized physicist or engineer to make quantitative sense of it. But simple dimensional reasoning accessible to any layman can tackle surprisingly sophisticated questions. With the energy, E, of the explosion an obvious parameter of interest, and also the density, ρ, of the air into which the fireball expands (a denser medium such as water may be expected to offer more resistance), to obtain the radius, R, of that fireball, which is a length, we have to form the dimension $[L]$ from those of E, ρ, and time, t. With some juggling of their dimensions expressed in $[L]$, $[T]$, and $[M]$ to cancel out all but the dimension $[L]$, we arrive at $R = k(Et^2/\rho)^{1/5}$. As in any dimensional argument, a dimensionless constant, the pure number k, is of course not fixed. Generally, these multiplicative constants have some small numerical value such as 1, 2, or π, and combinations thereof. As a result, all essential dependences and non-trivial scalings with energy and time are obtained without invoking any details of the complicated physics of an air explosion. An explosion with 10 times the energy released will expand to $10^{1/5} \approx 1.6$ times the radius in the same time. And an explosion at high altitude, where the density may be half of what is near the surface, will, all other parameters being equal, expand not to twice the size but rather only about 15% larger, the fifth root of 2 being about 1.15.

1.2 Adding a Dimension

With the above introduction to dimensions in physics, we turn now to the title theme of this chapter. It refers not to the introduction of new dimensions of time, mass, charge, etc., when needed, as in the previous section, but to the fact that, often in physics, it turns out to be useful to add a dimension to the problem under consideration as an aid to solving or better understanding it. This is purely as a device to calculate or understand better: no such extra dimension is physically present in the problem. This seems curious at first sight. One might think that introducing a new dimension (or a degree of freedom) to ones already present can only complicate matters. But time and again and in different areas of physics, we find instead that adding a dimension and, in particular, just one extra dimension simplifies the problem, even allowing an otherwise intractable problem to be solved. Often this is primarily a mathematical device (even that highlights its usefulness, now in mathematics) but it also gives in many instances greater insight into the physics involved.

Before turning to a simple example in physics, consider first a mathematical joke that illustrates this theme, with humour if not precision. A shepherd dies, leaving behind 11 sheep and a will stipulating that the eldest son is to inherit 1/2, the middle son 1/3, and the youngest boy 1/12 of the flock. As they puzzle over how to proceed, a wise shepherd uncle brings one sheep of his own that he adds to the flock, then gives six to the oldest son, four to the middle, and one to the youngest, and his own sheep is left behind for him to take away at the end. This admittedly simple trick does illustrate, however, the main theme of this chapter, of the merit in adding one element to solve an otherwise intractable problem. (For someone still puzzled by the uncle's sleight of hand/sheep involved, a moment's reflection on adding the fractions 1/2, 1/3, and 1/12 reveals the trick!)

1.2.1 *Linear Vibration to Circular Rotation: A Pedagogical Example*

One of the simplest motions in our physical world is that of simple harmonic oscillations of a mass point along a line, a one-dimensional motion in the direction x. The motion of the pendulum bob in the previous section or of a mass at the end of a spring (Figure 1.4) are examples that are usually presented in the very first lessons of physics. Two parameters characterize such simple harmonic motions, the frequency, ω, and the amplitude, A. The frequency is the inverse of the time period and, for a pendulum, is given by Eq. (1.1), $\omega = 2\pi/T = \sqrt{g/\ell}$. For a spring of elastic constant k, the frequency is $\omega = \sqrt{k/m}$, and so depends on the mass, m, stretching the spring. Ordinary language uses cycles per second (cps) or revolutions per minute (rpm). Each full

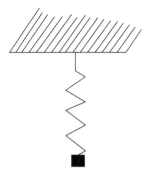

Figure 1.4 A simple spring.

cycle or revolution being 2π radians (360 in degrees), scientific usage is in radians/s, with units $Hz = s^{-1}$.

For the simplest case of fixed ω of any simple harmonic motion in one dimension, using the trigonometric function sin (or, alternatively, cos), we have $x(t) = A \sin \omega t$. It may seem surprising that this very simple motion with one parameter fixed and the instantaneous position varying in a simple fashion can be further simplified. But, if we adjoin another dimension, y, with a similar motion, $y(t) = A \cos \omega t$, then the combined two-dimensional motion is indeed simpler, with now both frequency and radius, A, fixed in time. All the 'complications', that the x and y positions vary with some mathematical (albeit simple trigonometric) dependences, are circumvented for the motion of the mass point on a circle, which is now at a fixed radius with a fixed rotational speed. Two fixed numbers, A and ω, describe the motion fully. The complications of the one-dimensional motion are then viewed as entirely due to the projection down in one dimension from the simpler uniform circular motion in a higher-dimensional space (Figure 1.5).

This is a very familiar example for many reasons and will recur in other themes of later chapters. One is for its connecting oscillations or vibrations with rotations. This also has an applied aspect for converting from one to the other, as in familiar examples of pistons in cars or trains (Figure 1.6). Note also the interesting interplay: the translational motion of the piston over a limited range (less than the size of a car or locomotive) converts to a rotational motion (rotations are always confined) and then to a translation of the whole car or train along the road or rail of unlimited range.

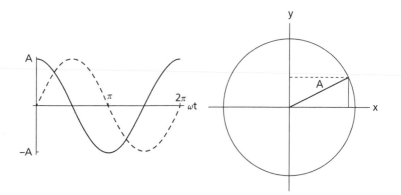

Figure 1.5 Simple harmonic motion on a line with amplitude A, and associated uniform motion in a circle of radius A. Shown on top are sine and cosine curves of the x and y projections of a point on the circle to the two Cartesian axes.

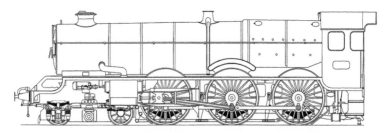

Figure 1.6 Pistons in an old steam railway engine that convert from linear to circular motion. 2007 Autocad drawing by Stavros1 of a Great Western King locomotive. <http://en.wikipedia.org/wiki/File:AutoCAD_drawing_of_a_Great_Western_King.png>

Yet another realization of the linear and rotational motions, shown in Figure 1.7, goes back to basics from the time of Galileo and Newton and is a classic problem in courses we teach in mechanics. A tunnel drilled through a diameter of the Earth (assumed to be a uniform–density sphere) leads to simple harmonic motion of a mass point dropped into it. The time period of this motion is that of a pendulum of length equal to the radius of the Earth, with a value as given by Eq. (1.1), $T = 2\pi\sqrt{6.4 \times 10^6/10} \approx 84$ min. This coincides with the time it takes a near-Earth satellite such as the International Space Station to go once around in a circular orbit (Figure 1.7). (Such a satellite is a few hundred miles above the surface but, with the relevant distance being from the

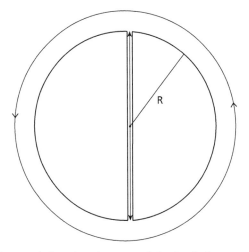

Figure 1.7 A tunnel though a diameter of the Earth (represented in cross-section by a circle) and a near-Earth satellite orbit. A mass point completes one full oscillation with the same time period of, approximately, 84 min for either motion, as given by the pendulum Eq. (1.1) with $\ell = R$, the radius of the Earth.

Earth's centre, the length involved is dominantly the Earth's radius of, approximately, 4,000 miles or 6,400 km.) These arguments are again illustrations of dimensional reasoning, the only relevant length being the radius of the Earth, which when coupled with the gravitational acceleration, g, provides a time through the pendulum equation Eq. (1.1). Naturally, the same time applies to the period of a (Gulliverian) pendulum the size of the Earth or to the time of free fall along a diameter or time period of a near-surface orbit, all these motions governed by g and the same distance, the radius of the Earth.

A closely related mathematical trick to the above addition of a dimension is familiar in ordinary integral calculus. (An integral is a generalization of addition, used to sum quantities that are continuously rather than discretely distributed.) One adjoins a second linear dimension to the first and goes to circular coordinates to simplify the integral $\int_{-\infty}^{\infty} \exp(-x^2)dx$. This 'Gaussian[6] integral' is converted by squaring it (since x is integrated over, it may be replaced by y in the second factor)

[6] Carl Friedrich Gauss, 1777–1855, German. One of the greatest mathematicians, with important contributions to physics and astronomy as well. A prodigious calculator who could handle complicated problems, he developed methods for computing celestial orbits, and wrote on electromagnetic phenomena.

into the simpler $2\pi \int_0^\infty \exp(-\rho^2)\rho\,d\rho$ because changing the variable to ρ with $\rho^2 = x^2 + y^2$ (together with the angle ϕ of circular coordinates; see Sec. 2.1) makes this a simpler, 'trivial' exponential integral, whose square root leads to the final desired result, $\sqrt{\pi}$.

As an example of a very sophisticated occurrence of the idea illustrated in Figure 1.5, the same figure appears (Figure 13.1 of [6]) in the construction of so-called maximally symmetric spaces in differential geometry, especially as used in Einstein's General Theory of Relativity to describe gravitation. Again, the details are not important for the point being made here but Einstein's theory is one of geometry, the geometry of space–time (Sec. 1.2.4). In analysing curved surfaces and the steps of differential calculus involved in the differential geometry of them, spaces satisfying certain specified symmetries (see Chapter 5) are of interest. Using the horizontal line in Figure 1.5 as a stand-in for a complicated non-Euclidean N-dimensional space, and embedding it in a simpler, flat $(N + 1)$-dimensional space represented by the circle, with an extra dimension as the vertical direction, permits the desired construction. Even more than in the simple example presented in Figure 1.5, here one can truly appreciate the astonishing power of such embeddings in an extra dimension, all the complicated curvature of the N-dimensional space contained in that constraint of coming down in one from the larger (but simpler) dimensional space.

But the simple example in Figure 1.5 already illustrates the underlying theme, that dynamics can be subsumed into kinematics in one dimension higher and then all the complicated and non-trivial dynamics are realized in the kinematical constraint that reduces the problem to the lower dimension. The following sub-sections will consider increasingly sophisticated illustrations of the theme and are aimed at physics students familiar with the basics of calculus and quantum mechanics.

1.2.2 Green's Theorems

In ordinary single-variable calculus, integration and differentiation as inverses of each other lead to the simple result:

$$\int_a^b \frac{df(x)}{dx}dx = f(b) - f(a). \tag{1.2}$$

The extension by George Green[7] to two-variable functions, $P(x, y)$ and $Q(x, y)$,

$$\int \left(\frac{\partial Q}{\partial x} - \frac{\partial P}{\partial y} \right) dx dy = \int_C (Q dy - P dx), \qquad (1.3)$$

where the right-hand line integral is over the closed contour, C, around the area of the left-hand integral, is perhaps one of the most important theorems of mathematical physics. Many other theorems, carrying the names of Gauss, Stokes[8], divergence, etc., are special cases and occur throughout physics. All share the feature that an integral over some dimension is equivalent to another over the next larger dimension. Our starting Eq. (1.2) already contains the essence in that the integral on the left-hand side is a sum of the values of the integrand over a whole one-dimensional line interval but, on the right-hand side, reduces just to the difference in values at the two end points, amounting to a step down to zero dimensions.

Physics deals throughout with 'vectors', quantities that require both a magnitude and a direction to specify them, examples being position, velocity, acceleration, electric and magnetic fields, etc. Since variations in them also have a directional sense, a change in the x direction being generally different from that in the y or z directions, the differential operation of calculus describing variations also acquires a vectorial characteristic, differentials in the three orthogonal directions lumped together into the vector differential operator ∇ : $(\partial/\partial x, \partial/\partial y, \partial/\partial z)$. From any two vector quantities $\vec{A}$ and $\vec{B}$, one can form products of two kinds, depending on whether one ends with a magnitude alone, called a 'scalar', this being the 'scalar product', denoted with a dot as $\vec{A} \cdot \vec{B}$,

---

[7] George Green, 1793–1841, English. A miller by day and self-taught mathematician who wrote an astonishing essay in 1828, providing a mathematical analysis for electricity and magnetism, in which he introduced the theorems and functions now named for him and which find wide applications throughout physics and engineering. The essay was finally recognised and he went to Cambridge University, obtained a degree, and became a fellow but, unfortunately, became ill and died relatively young.

[8] George Gabriel Stokes, 1819–1903, Irish. Professor of mathematics at Cambridge University for over 50 years, who made important contributions to mathematical physics, optics, and fluid dynamics (including 'Stokes law' for the viscous friction on a sphere in a fluid medium). The Stokes theorem is said to have originated with fellow professor Kelvin, who suggested it in a letter, Stokes then using it in a prize examination and his name becoming attached to it. Maxwell, among others, became aware of it from there.

or another vector, called the 'vector product', denoted with a cross as $\vec{A} \times \vec{B}$. This is also true when the vector differential, ∇, acts on a general vector $\vec{V}$, the scalar 'divergence', $\nabla \cdot \vec{V}$, and the vector 'curl', $\nabla \times \vec{V}$, being important concepts throughout physics. Various physical significance can be associated with them, immediate ones, as suggested by their names, being a spreading (or increase/decrease) or a twist (or rotation), respectively. The theme of adding a dimension enters naturally with these vector differentials and is familiar to every physics student, even if it is not always recognized as such. 'Gauss's law' or the 'divergence theorem' relates the volume integral (over $d\tau$) over the divergence of a vector field, $\nabla \cdot \vec{V}$, to the surface integral (over $d\vec{\sigma}$) of that field, $\vec{V}$, over the surface bounding that volume:

$$\int \nabla \cdot \vec{V} d\tau = \int \vec{V} \cdot d\vec{\sigma}. \qquad (1.4)$$

Newton's law of gravitation, expressed as the flux of the gravitational field due to masses contained within the volume, or the similar Coulomb's law for an electric field due to electric charges, involves this expression applicable to any vector quantity in physics. The surface integral over the field is simply the total charge or mass contained in the inside volume. The closed surface (two dimensional) bounding the volume (three dimensional) is a higher-dimensional realization of the two end-points (zero dimensional) bounding the line integral (one dimensional) in Eq. (1.2).

Stokes's law, relating the integral of the curl of a vector over an area to the line integral (over $d\vec{\ell}$) of the vector along the boundary of that area, is another example:

$$\int \nabla \times \vec{V} \cdot d\vec{\sigma} = \int \vec{V} \cdot d\vec{\ell}. \qquad (1.5)$$

And, as an illustration of a result well known in multi-dimensional geometry, with the choice of $\vec{V} = \vec{r}$, the radial vector in any d-dimensional space, since $\nabla \cdot \vec{r} = d$, Eq. (1.4) relates the volume of a sphere in d dimensions to its surface area, again familiar already from the $4\pi/3$ and 4π, respectively, of school geometry as the multiplicative factors for the volume and surface area of a sphere of radius a in three dimensions.

Besides their ubiquitous appearance throughout all areas of physics, such Green's theorems have a profound philosophical depth. Information contained in an entire volume, even perhaps of the whole Universe, can be viewed equivalently from that available on the surface at large distances. While much is made in recent times of so-called holographic principles in general relativity or string theories [7], this idea is not new, as we saw in elementary mathematics or in physics. After all, we study distant stars and galaxies all the way to the Big Bang of the earliest moments of our Universe, all that content filling the time history of the Universe, through the information now available to us on a small patch of the surface that we occupy where the Universe has expanded to in our times.

In dealing with quantities that are neither created nor destroyed, 'conserved' in physics terminology, whether the mass of a fluid in flow or the amount of charge in motion (a current), or the number of cows in a stockade, if one keeps track of the amount entering or leaving, that amount must be compensated by a corresponding equal change in the amount or number contained within. This is a law of conservation (see Sec. 5.1.2 for their connection to symmetries). The differential equation expressing this statement, called an 'equation of continuity', and an associated Green's theorem, are of great importance throughout all areas of physics.

Quantum physics, in particular, has made use of this idea from its earliest days. Unlike in classical physics with positions and velocities of mass points, the physical state of a system is described in quantum physics by a complex (in the mathematical sense of involving the imaginary unit $i = \sqrt{-1}$) 'wave function', ψ. The probability interpretation, first given by Max Born[9], that the modulus squared, $|\psi|^2$, represents the probability of finding the system in an interval about the corresponding variable, whether position, momentum, or anything else, requires

---

[9] Max Born, 1882–1970, German and British. A founder of quantum mechanics and the originator of the probability interpretation of the wave function. Together with his assistant Pascual Jordan, he rendered his student Heisenberg's discovery into matrix language and wrote the basic commutator between position and momentum that underlies Heisenberg uncertainty principle. Born also developed, with Robert Oppenheimer, an approximation technique for molecular structure that continues to dominate that field. Another widely used method in scattering theory bears his name as well. And he made important contributions to our understanding of crystal lattices and optics.

that $|\psi|^2$ be 'normalized' to unity (or, equivalently, to a Dirac[10] delta
function for continuous distributions). This is the statement that $|\psi|^2$,
when integrated over the full space, must yield the unity expected of
the total probability that the system is somewhere. But throughout
microscopic physics, whether of atoms, nuclei, or elementary particles,
most of this space is inaccessible to our measuring apparatus. And this
is where Green's theorems allow us to use just the knowledge of the
wave function and its first derivatives at the surface at infinity, which
are accessible to our laboratories, to accomplish this normalization (or
get any of the so-called scattering parameters when one particle scatters
off another, all of which are defined only at infinity; see Sec. 7.4). It also
fits into the philosophy, especially emphasized by Bohr[11] and Heisenberg[12], of using in physics only constructs that are at least in principle
accessible to our measuring apparatus (Sec. 7.4).

[10] Paul A. M. Dirac, 1902–1984, English. One of the three founders of quantum
mechanics, his transformation theory and bra-ket notation are now standard in physics.
Made quantum mechanics compatible with Special Relativity through the 'Dirac equation', which incorporated quantum-mechanical spin angular momentum and, further,
predicted the existence of anti-particles to electrons, protons, etc. He took the fundamental steps towards quantum field theory and to the variational or path integral
approach to quantum physics. Also, he introduced the distribution function that bears
his name and is used extensively in physics for continuous distributions.

[11] Niels Bohr, 1885–1962, Danish. Principal founder of quantum physics when in 1913
he used Planck's idea of the quantum and constant $\hbar$ to account for the structure of
the hydrogen atom based on Rutherford's experiments and the empirical formula of
Balmer for the line spectra of the hydrogen atom. He extended these ideas to higher
atoms but, even more significantly, guided the development of quantum mechanics
through a unique school of physics he headed in Copenhagen, hosting most of the
prominent quantum physicists of the day. He also shaped the philosophy of the subject,
through his debates with Einstein and through his formulation of the Correspondence
Principle and complementarity. Following the discovery of nuclear fission, he developed the liquid drop model of nuclear structure that accounts for fission of nuclei
by neutrons. He served as a consultant to the Manhattan Project, which developed
nuclear weapons and fission energy.

[12] Werner Heisenberg, 1901–1976, German. His papers in 1925 launched the subject
of quantum mechanics in the form called matrix mechanics. His uncertainty principle
distilled the role of complementary observables in quantum physics. He also made pi-
oneering and important contributions to dispersion relations and scattering theory,
nuclear physics, ferromagnetism, and elementary particle physics. During World War
II, he headed an unsuccessful German effort to develop nuclear reactors, and after that
war he played a role in rebuilding German physics institutions, notably the Max Planck
institutes.

Indeed, normalization is an aspect of 'unitarity' or conservation. In the preceding paragraph, it is the conservation of probability. In the previous one, it was the conservation of fluid mass, or charge, or number of cows. Conservation laws in physics are expressed through equations of continuity that go even further back and are important also in classical physics. They express this philosophy that allows us to keep track of changes in time of some physical quantity over an entire volume simply through observations of the flux through the bounding surface, often at large distances from the centre. When that physical quantity is conserved, the flux through the surface must be balanced by sources or sinks within the volume.

1.2.3 Lagrange Multipliers for Extremum Problems

Locating maxima and minima of functions or physical quantities is often of interest in mathematics, science, and engineering. The shortest path or shortest time for covering it, the lowest energy, the maximum efficiency, and many such extremum questions arise very commonly. For a function of one variable, Newton's differential calculus provides the solution. The derivative of the function vanishes at such extreme points and serves to locate them and determine the desired value there. More complicated prescriptions are necessary for many-variable calculus, and one also refers to them more generally as stationary points because, in addition to overall peaks and valleys, other situations now arise, such as saddle points, a theme to be taken up in Chapter 3. The problem becomes more complicated when constraints are specified under which this determination is to be made. The method of Lagrange undetermined multipliers for locating maxima or minima or, more generally, stationary points, whether of one- or many-variable calculus when constraints are present, is another familiar example of adding a dimension, even if it is not usually presented as such.

It works also in the calculus of variations, a subject that deals not just with functions of algebraic variables but also with whole functions themselves [8]. Such 'functionals' over functions are extremized subject to constraining equations that are often part of the definition of the problem. A classic problem of the ancients was to determine the maximum area enclosed for a given perimeter. (The answer is a circle.) An especially simple and commonplace example is of finding the shortest path between A and B, for instance on the surface of the Earth. It is formulated by asking of all possible paths (functions), which one has the least length, that length being the functional and depending on the path between A and B. Thus the geodesic or shortest path on a general surface (such as great circles on our globe) is viewed as extremizing the expression for metric distance while regarding the defining equation of that particular surface as a constraint. The path is constrained to lie at all times on the surface. Ever since Lagrange's[13] reformulation of Newtonian mechanics as an extremum principle, and going even further back to Fermat's[14] principle of least time (for the path of light between A and B with material of varying refractive index in between) that predates and anticipated Newtonian calculus, such formulations of the basic laws of motion in physics have proved powerful, both in practical use and in providing insight into the structure of our subject [9].

After mechanics and optics, quantum mechanics, quantum field theories, and general relativity have all been cast as stationary principles. To be discussed more in later chapters, these are discussed in terms of a 'Lagrangian', typically in mechanics the difference in kinetic and potential energies of the physical system. Indeed, with 'action', defined as the integral of the Lagrangian over time or of the Lagrangian (space) density over space and time, having the same dimensions as that of Planck's[15]

---

[13] Joseph Louis Lagrange, 1736–1813, French. One of the most eminent mathematicians and astronomers of the 18th century. He recast Newton's equations in advanced calculus, computed planetary orbits, and contributed to mathematical analysis (especially the calculus of variations) and number theory. Lagrange points in the three-body system with the Sun and a planet (Sec. 3.1.1), where natural and artificial satellites can be parked, are named for him. He played a dominant role in adopting the metric system in France, was a senator, and headed the Bureau of Longitude in Paris.

[14] Pierre Fermat, 1601–1665, French. Lawyer and mathematician with deep contributions to analysis and number theory. His principle of least time for light propagation is the earliest example of the calculus of variations and, with his discussion of limits and tangents, he anticipated the infinitesimal calculus of Newton and Leibnitz. His 'last theorem' on sums of higher-than-two powers of integers became one of the most recognized unsolved problems of mathematics for centuries, finally solved in 1994. In an exchange of letters with his friend Pascal, he enunciated the basics of probability theory.

[15] Max Planck, 1858–1947, German. An expert in thermodynamics, in trying to understand black-body radiation when the radiator and the radiation emitted are in equilibrium, he introduced a new way of counting energy in discrete quanta to account for the experimental observations of his colleagues. Later, through the extensions and elaborations by Einstein and Bohr, this gave rise to quantum physics, although Planck himself never accepted its main philosophical elements. Planck saw the fundamental constants $\hbar$ and k, now named for Planck and Boltzmann, respectively, as a crucial contribution towards defining universal measures for physical quantities, and the Planck length, Planck time, and Planck mass are now familiar, especially in cosmology and

quantum constant $\hbar$, the stationary action principle is in some ways the most natural formulation of quantum physics. Feynman's[16] path integral formulation puts the action, S, along a path divided by $\hbar$ in the exponent as $\exp(iS/\hbar)$ and the quantum wave function ψ (see Sec. 1.2.2) is recovered as the sum over all possible paths (Sec. 7.2). The classical limit of quantum physics also becomes natural in this language as the situation when the classical path or orbit dominates that sum (see also Sec. 8.5).

For our purposes here of adding a dimension as an illustration of the use and role of Lagrange multipliers, consider a simple problem that often occurs in undergraduate courses. Suppose the temperature in three-dimensional space varies from one location to another according to $T(x, y, z) = xyz$; that is, the temperature is given by the product of the coordinate values of the point. If we wish to locate the points on a sphere of radius R where T is a maximum and to find the value of that maximum, we have here a typical problem of multi-dimensional (here three-dimensional) calculus. A straightforward approach would express one of the variables in terms of the other two through the defining equation of the sphere, $x^2 + y^2 + z^2 = R^2$, and maximize the remaining expression for temperature as a function of two variables.

gravitation. Planck was a founding member of the German Physical Society and shepherded it and physics institutions through the Nazi era, with the hope of preserving them through it and rebuilding German physics after the war. Today, the string of German physics research institutes bears his name.

[16] Richard P. Feynman, 1918–1988, American. An extraordinary theoretical physicist, one of the co-formulators of the first relativistic quantum field theory, called quantum electrodynamics (QED). He invented a unique diagrammatic technique that is universally used in the field theories of today. A master of variational techniques, he followed early work by Dirac in formulating a path integral approach to quantum mechanics and, with his teacher John Wheeler, to radiation. He also made fundamental contributions to superfluidity and elementary particle physics. He was a member of the theory division of the Manhattan Project, which developed the nuclear bomb, where he used early computers for numerical calculations. A gifted teacher and expositor, he achieved legendary status among contemporary physicists through a course of lectures covering all physics that, in its three volumes, has educated and influenced physicists around the world. A slim volume, QED [3], and scores of his writings and books, many displaying his playful humour and style and some appearing posthumously, and his prominent role in the national commission that investigated a space shuttle launch disaster, made him a household name. An essay of his on the exploitation of microscopic devices is seen as the inspiration of today's field of quantum computing and information.

However, this would involve complicated equations with square roots and differentials of them.

Instead, Lagrange's method of 'undetermined multipliers' introduces a functional,

$$\hat{T}(x, y, z, \lambda) = xyz - \lambda(x^2 + y^2 + z^2 - R^2), \qquad (1.6)$$

where we introduce a new parameter (or dimension), λ, that is at this point undetermined and which multiplies the constraint 'as a zero'. Whatever the value of λ, $\hat{T}$ coincides with the T we seek for the problem posed, the added term vanishing because of the zero multiplying λ. Again, while it may seem that one is complicating a three-dimensional problem further by going to four dimensions with a new variable, the important point is that we can now proceed to handle $\hat{T}$ as unconstrained, freely varying it with respect to all four variables (x, y, z, λ) to seek the stationary point. It is clear from Eq. (1.6) that λ has dimensions of length. All four dimensions, including the introduced, 'fictitious', λ, are viewed on an equal footing and as independent. The four equations, one being just the constraint equation itself, are then solved simultaneously. In problems such as this with an underlying symmetry (see Sec. 5.1.1), even this step proves trivial and one is led to the point $x = y = z = R/\sqrt{3}$, together with its three 'symmetrical partners', with two of the coordinates involving the negative value of the square root, as the points of maximum temperature with value $T = R^3/3\sqrt{3}$.

This illustrates the central theme of this chapter, that adding a dimension, here one for each constraint, can simplify the calculation and our understanding of the physics involved. While the final value of λ is itself irrelevant ($\lambda = R/2\sqrt{3}$ in this example), in that it multiplies a zero in Eq. (1.6), it may be regarded as a 'force' enforcing the constraint. Once again, there is a nice philosophical element, that what seem to be forces governing dynamics in space may be viewed equivalently as kinematics in the space with one more dimension together with a kinematical constraint (one more dimension and constraint with each Lagrange multiplier), exactly as in the complicated example considered at the end of Sec. 1.2.1.

The idea of adding a zero to incorporate the laws of physics or the defining equations involved to calculate some other physical property of the system allows a general construction of stationary or variational principles for any such property [10]. The Lagrange multipliers may not always be numbers as in the present example but may be functions, matrices, etc., as appropriate in constructing the extended functional expression for the property of interest.

1.2.4 Space–Time

One canonical example of adding a dimension, now even familiar to the man on the street, is the view of the world in space–time. Mechanics, having to do with changes in time of three-dimensional location, has always had these four dimensions, three of space and one of time, as its natural stage. Following Einstein's Special Theory of Relativity, physics has recognized that they are indeed intertwined.

An important consideration in physics is of objects that remain unchanged under some transformation of the coordinate axes. Such transformation symmetries and their corresponding 'invariants' are the topic of Chapter 5, but consider here the distance $(x^2 + y^2 + z^2)^{1/2}$ of the point (x, y, z) from the origin as an invariant under translations and rotations of the axes. The three coordinates assigned to the point and to the origin may change under such translations or rotations of the coordinate axes being used to describe them, but the distance of separation remains unchanged. In Special Relativity, it is replaced by the invariant interval or, equivalently, its square $(c^2t^2 - x^2 - y^2 - z^2)$. The space–time interval involves all four dimensions of space and time. At the same time, instead of the three-dimensional vector $\vec{r}$ with components (x, y, z), space and time together as a four-component vector (often simply called a four-vector) $(ct, \vec{r})$, along with similar four-vectors of energy-momentum $(E/c, \vec{p})$, provide the natural language for relativistic mechanics. Note a change in sign of the space components relative to the one of time in the space–time interval, which reflects the differing status of space and time even when combined into a set of four. And, note of course, the dimensional aspect that uses c, the speed of light, to put space and time, or energy and momentum, on an equal dimensional footing in their four-vectors.

Quantities that do not change under rotation of the axes, such as the distance of separation in the previous paragraph, are called 'scalars'. Indeed, unlike the looser description earlier as something with magnitude alone and no directional sense, this is the proper definition of what is meant by a scalar in physics: it is an object left unchanged by rotations. Mass, charge, temperature, etc., are other examples, all taking some numerical value in whatever system of units is used to describe them, but remaining the same regardless of a particular set of orthogonal axes in consideration or a second set, usually called primed and so designated by using primes on x, y, and z, obtained through a rotation of the axes. On the other hand, a directed line segment that has both a magnitude for the distance of separation and a direction of where the second point lies with respect to the first, is a 'vector', with a definite relation between the values of (x', y', z') and (x, y, z) depending on the rotation involved. Besides this basic vector quantity, any other set of three components is called a vector if the same relation exists between its primed and unprimed components. Velocity, acceleration, an electric and a magnetic field are all examples of such vector quantities. They all transform in like manner and like the basic vector $\vec{r}$: (x, y, z). (Note a philosophical theme of central importance, to be developed further in Chapter 2, of using behaviour under certain transformations to define objects in physics.)

For notational convenience, an index notation is adopted, with r_i standing for x, y, and z when i equals 1, 2, and 3, respectively. Similarly a_i for acceleration and B_i for magnetic field denote the components of those objects along the three directions. The index notation allows easy extension to descriptions of any number of dimensions, i running over 1, 2, 3, and 4 for instance in a four-dimensional world. It also allows an easy definition of other objects, called 'tensors', of higher rank, again according to a well-defined connection between their primed and unprimed components that is given once and for all for a specified rotation, independent of the object under consideration. Just as any vector, whether velocity or electric field, transforms in the same way, so does any tensor of a particular rank with a well-defined relationship between its primed and unprimed components, depending on the rank. That is how the tensorial nature of a physical quantity is defined.

Scalars are said to be of rank zero, in that they do not change at all, while vectors are said to be of rank one, and an example of a higher-rank tensor is the moment of inertia, which has rank two. It expresses the relation between the angular momentum of a rigid body, itself a vector or tensor of rank one, and angular velocity, also a vector object. For a complex object, the two vectors of angular momentum, $\vec{\ell}$, and angular velocity, $\vec{\omega}$, are not always simply related through a (scalar) multiplicative constant but a change in angular velocity in one direction can cause a change in angular momentum in a different direction. This is expressed through $\ell_i = I_{ij}\omega_j$, with I the moment of

$$\begin{pmatrix} I_{11} & I_{12} & I_{13} \\ I_{21} & I_{22} & I_{23} \\ I_{31} & I_{32} & I_{33} \end{pmatrix}$$

Figure 1.8 A 3×3 matrix representing the second-rank tensor of moment of inertia. The elements I_{ij}, with $i, j = 1, 2, 3$, relate the angular velocity in any of the three axes directions j to the angular momentum components i. With $I_{ij} = I_{ji}$, this is a symmetric array, elements equal under reflection about the diagonal so that there are only six independent numbers in the array.

inertia tensor, and an assumed convention that when an index is repeated, there is an implied summation over all values of that index. Thus, $\ell_1 = I_{11}\omega_1 + I_{12}\omega_2 + I_{13}\omega_3$, and there are two similar equations for ℓ_2 and ℓ_3, so that in general any of the three components of angular velocity can affect a particular component of angular momentum. The coefficients of this proportionality constitute the moment of inertia tensor.

With two indices (i, j), each taking three values in three dimensions, there are nine components in general for a tensor of rank two that can be conveniently represented as a 3×3 square matrix, although there is often a reduction in number because of the symmetry properties of the object in question (Figure 1.8). Thus, the moment of inertia tensor is symmetric, interchange of i and j not changing the physics, and there are only six independent components, three along the diagonal and three off-diagonal elements of the matrix which are repeated across the diagonal. Further reduction in the number of components signifies further geometrical symmetries of the body. On the other hand, an antisymmetric tensor of rank two, that is one that changes sign if i and j are interchanged, will have only three non-zero components, the diagonal elements being necessarily zero and the off-diagonal ones related by a minus sign to corresponding elements reflected through the diagonal. In a d-dimensional world, clearly a vector has d components, a symmetric rank two tensor has $d(d + 1)/2$, and an antisymmetric rank two tensor $d(d - 1)/2$ components.

Electric ($\vec{E}$) and magnetic ($\vec{B}$) fields are three-dimensional vectors, each with three components. These six components of electric and magnetic field vectors group together (Figure 1.9) naturally into a single object, an antisymmetric tensor of rank two in four dimensions, with well-defined behaviour under the so-called Lorentz[17] transformations of Special Relativity. Also, the vector potential, $\vec{A}$, and scalar potential, Φ, from which the fields are derived,

$$\vec{E} = -(1/c)(\partial \vec{A}/\partial t) - \nabla \Phi, \quad \vec{B} = \nabla \times \vec{A}, \qquad (1.7)$$

transform as a four-vector $(\Phi, \vec{A})$.

Both mechanics and electromagnetism are unified in this four-dimensional view with the important invariant being a speed, that of light, c, speed of course combining both space and time. Lorentz transformations include, besides rotations of spatial axes, velocity 'boosts' connecting two frames moving with uniform velocity with respect to one another. Such frames are called 'inertial frames' and the central tenet of Special Relativity is that physics remains invariant under changes from one such inertial frame to another, that all such inertial frames are on a par as regards the laws of physics. Absolute rest and absolute velocity have no physical significance, only relative velocities, and, for physics to have meaning, no inertial frame is on a special footing. This was already evident to Galileo and Newton for mechanics, and Einstein's Special Theory of Relativity extends it to all physics. In doing so, it transformed our understanding of space and time.

Another crucial precursor for Einstein was the work of Maxwell, who himself built his equations of electric and magnetic fields (Figure 1.10) based on the important concept of a field, which was introduced into physics by Faraday[18]. Unlike Newton's gravity, which was a force or interaction that acted 'at a distance' between

[17] Hendrik Antoon Lorentz, 1853–1928, Dutch. Theoretical physicist who studied electricity, magnetism, and mechanics. He developed a theory of magnetic field effects on atomic spectra to explain the Zeeman effect. In studying the motion of charged particles (the force law carries his name), he formulated the 'Lorentz transformations' without a full understanding of them, which was provided by Einstein's Special Relativity. He chaired the first Solvay Conference, which, together with later ones in that series, brought together all the prominent physicists of the day and was very influential in the development of early 20th-century physics.

[18] Michael Faraday, 1791–1867, English. Physicist and chemist, mostly self-taught and a laboratory assistant to Humphrey Davy, whom he succeeded in the Royal Laboratory. His physical picture of magnetic fields and of the laws of induction inspired Maxwell in formulating the laws of induction and the connections between electricity and magnetism. The unit of electrical capacitance is named the farad.

$$
\begin{pmatrix}
0 & E_x & E_y & E_z \\
-E_x & 0 & B_z & -B_y \\
-E_y & -B_z & 0 & B_x \\
-E_z & B_y & -B_x & 0
\end{pmatrix}
$$

Figure 1.9 The second-rank electromagnetic tensor, $F_{\mu\nu}$, with $\mu, \nu = 0, 1, 2, 3$, arranged in an antisymmetric (with respect to reflection about the diagonal) 4×4 matrix array of electric (E) and magnetic (B) field components.

$$
\nabla \times \vec{E} = -\frac{1}{c}\frac{\partial \vec{B}}{\partial t}
$$

$$
\nabla \cdot \vec{B} = 0
$$

$$
\nabla \times \vec{B} = \frac{1}{c}\frac{\partial \vec{E}}{\partial t} + \frac{4\pi}{c}\vec{j}
$$

$$
\nabla \cdot \vec{E} = 4\pi\rho
$$

Figure 1.10 Maxwell's equations relating electric and magnetic fields in vector form, written as two sets of equation pairs, each themselves one scalar and one vector equation.

two bodies, however far apart, the casting in terms of a field defined everywhere, including at locations between the bodies, was crucial in the development of physics. A mass or an electric charge sets up a corresponding gravitational or electromagnetic field around it, and another mass or charge reacts to it. With the fields themselves capable of interacting between neighbouring points in space and time, and capable of carrying energy, momentum, etc., the description of an interaction can be made a local one.

The occurrence of c, the speed of light (a historically much earlier concept), intrinsically in the equations governing electric and magnetic fields, even in vacuum, gives significance to a natural speed (the same as seen by any inertial frame) for even the establishment of a field around a mass or charge. It is with this finite speed that the presence of such a source can be felt in the region around it, a distant mass or charge sensing it only after a finite time. Light itself, whether in the visible or other ranges of the spectrum, is then seen by all inertial observers as such propagating waves with speed c (in vacuum) of electric and magnetic fields. (Gravitational waves, yet to be directly observed, will also travel with the same speed.)

The difference between rotations about the three spatial axes and velocity boosts along each of the axes goes hand in hand with the different status of the time axis relative to the others that was noted in the change in sign in the space–time interval. Time is not just a fourth coordinate axis. Even in the mathematics of handling four dimensions together, this distinction between space and time is worth keeping in mind and is expressed by saying that the 'metric' for measuring lengths has opposite signs between the two. Correspondingly, the square root of (-1) being the imaginary unit i, trigonometric functions of sine and cosine in rotation angles pass into hyperbolic functions sinh and cosh for the velocity boosts of Lorentz transformations. This is the basis for saying that time is a fourth coordinate but an imaginary one!

Just as three-dimensional rotations transform a vector's components (x, y, z) to (x', y', z') in the rotated frame, with the squared length an invariant, the larger set of Lorentz transformations that include velocity boosts transform between four-vectors in inertial frames. The counterpart of the Lorentz scalar space–time interval, $\sqrt{c^2 t^2 - x^2 - y^2 - z^2}$, as an invariant is the similar one for energy-momentum, $\sqrt{E^2 - c^2(\vec{p})^2}$, which is identified as mc^2, the rest mass energy of the particle. This equivalence of mass and energy is one of the central results of Einstein's Special Theory of Relativity.

The tenets of Special Relativity, and the equivalence of inertial frames under all Lorentz transformations that include not only rotations but also Lorentz boosts, make the four-dimensional space–time the natural framework for physics. Under rotations alone, electric and magnetic fields are not mixed up (only their components are among themselves) but they are by boosts (what may appear as a pure electric field in one inertial frame may be both electric and magnetic fields in another, with $(E^2 + B^2)$ and $\vec{E} \cdot \vec{B}$ invariant) so that they are indeed to be seen as the six components, shown in Figure 1.9, of an antisymmetric tensor of rank two. All four of Maxwell's equations in three-vector language shown in Figure 1.10 can then be rendered compactly in four-vector language in Figure 1.11 along with the definition from Eq. (1.7) of the fields in terms of the vector potential, $\vec{A}$, and scalar potential, Φ, or, together, the four-potential $A_\mu : (\Phi, \vec{A})$; the four-current is $j_\mu : (c\rho, \vec{j})$. Because of time and space not being entirely on a par, it is customary to use in place of Latin indices i and j running over four values, the Greek μ

and ν, taking the four values 0, 1, 2, and 3, for describing space–time in physics, the first index 0 denoting time.

The first two Maxwell's equations in Figure 1.10 are naturally subsumed in the definition of the antisymmetric tensor of the fields from the four-potential,

$$F_{\mu\nu} = \partial_\mu A_\nu - \partial_\nu A_\mu, \qquad (1.8)$$

where ∂ is now the four-dimensional differential operator in $(ct, \vec{r})$. This is another elegant simplification in recasting electromagnetism from fields in terms of potentials, these forming a four-vector and also incorporating the first two of Maxwell's equations (actually four equations, one being a vector equation) into the rules of calculus obeyed by the differential operator ∇.

The recasting of the larger and more complex set of Maxwell's equations in three-vector language in Figure 1.10 into the elegantly compact four-vector form in Figure 1.11 is, in part, one of notation, including implied summation of a repeated index and the properties of the differential $\partial_\mu : ((1/c)\partial/\partial t, \nabla)$. The merit of this four-dimensional view is that it also makes obvious not only the invariance under Lorentz transformations because of like transformations of vectors and tensors on two sides of an equation, but, further, again as per the rules of calculus, also the equation of continuity that follows upon further differentiation: $\partial_\mu j_\mu = 0$ (obvious because of summation over all μ and ν with $\partial_\mu \partial_\nu$ symmetric and $F_{\mu\nu}$ antisymmetric under interchange of (μ, ν)), which is an expression of the conservation of electric charge (see Sec. 5.2.2). This is a nice illustration, both of this chapter's theme of adding a dimension and of another theme that occurs elsewhere in this book (Sec. 2.3) of the power of a notation that seems so natural a fit to the physics of our world.

Einstein's General Theory of Relativity goes further in considering even more general transformations of the coordinates than the Lorentz

$$\partial_\nu F_{\mu\nu} = \frac{4\pi}{c} j_\mu$$

Figure 1.11 Maxwell's equations in four-vector form, incorporating all equations in Figure 1.10 into an equation for the electromagnetic tensor in Figure 1.9 and the four-vector current, j.

transformations of the Special Theory (see Sec. 5.2.6). It is based on the symmetric 'metric tensor' $g_{\mu\nu}$, $\mu, \nu = 0, 1, 2, 3$, with its 10 independent coefficients (antisymmetric tensors of the previous paragraph involve $4 \times 3/2 = 6$ and symmetric ones $4 \times 5/2 = 10$ elements). Interestingly, adding one more dimension, a fifth, as Kaluza[19] and Klein[20] did, did not lead to something of value. Although at first sight attractive, the $5 \times 6/2 = 15$ components of a symmetric tensor of rank two now possibly accommodating together Einstein's General Relativity equations with Maxwell's in an attempt to unify gravitation and electromagnetism, did not, however, prove fruitful as physics.

1.2.5 The Hydrogen Atom

The aspect of an added dimension also characterizes microscopic entities such as atoms and, in particular, already the simplest of them, the hydrogen atom. Before turning to the additional dimension, we consider first some elements of atomic structure. Ever since the Greeks and other ancients, the idea that there is an elementary or smallest component of all matter, that matter's divisibility has a limit in constituent atoms, has been a powerful principle, one also in physics. It was already clear over 100 years ago, from the gross properties of matter, that the size of atoms is small on our everyday scale. The 1,000-fold difference in density of water vapour from that of water as liquid or water as ice pointed to individual atoms or molecules being 10 times (again, a dimensional element, the cube root of 1,000) further apart than their size in the former gaseous phase while bumping against each other in the two condensed phases. And, from the gas constant relating the gross thermodynamic quantities of pressure, volume, and temperature, a size of about 10^{-10} m or 0.1 nm could be ascribed to individual atoms.

Next, almost exactly 100 years ago, Rutherford[21] discovered that nearly all the mass of an atom is concentrated in a positively charged

[19] T. Kaluza, 1885–1954, German. Mathematician and physicist, and learned in several languages. He discovered that writing Einstein's equations in five dimensions gave a natural way to embrace Maxwell's equations as well.

[20] Oskar Klein, 1894–1977, Swedish. Theoretical physicist who had the idea that extra dimensions may be real. His name is also attached, along with that of Gordon, to a relativistic quantum field theoretic equation for spinless particles (Sec. 7.3).

[21] Ernest Rutherford, 1871–1937, British. Pioneer nuclear experimentalist who studied various forms of radioactivity and transformations between nuclei. Most known for his planetary model of atoms, with electrons in orbit around a smaller, central,

nucleus with much lighter negatively charged electrons around it, the size of the nucleus being 100,000 times smaller than the atom as a whole. This was the only explanation for his experimental observations that tiny positively charged alpha particles, when fired at metal foils, were sometimes reflected almost backwards. Only a very tightly confined positive charge, confined to much less than 0.1 nm, could have repelled them so.

The lightest and simplest atom, that of hydrogen, has just one electron around a proton that is its nucleus, the electron being nearly 2,000 times less massive than the proton. They being equally and oppositely charged electrically, the atom as a whole is neutral but bound by their electrical attraction. The helium atom has two electrons and its nucleus has two protons and two neutrons, the latter of almost the same mass as the proton but electrically neutral. And so on, through the Periodic Table of naturally occurring elements (higher elements have also been created artificially) up to uranium, with 92 electrons around a nucleus that has 92 protons and a much larger number of neutrons, varying in number with the particular isotope of uranium (hydrogen also has rarer isotopes with one and two neutrons in addition to the proton, and helium too has a rarer isotope that has two protons but only one neutron in its nucleus, and exotic, radioactive species with several neutrons in a 'halo' are also created in some laboratory experiments).

Understanding hydrogen's structure is the key to understanding all atoms and matter, and it was clear also that there was a major problem in doing so in terms of classical physics. If the electrons are orbiting the nucleus (somewhat as in a Solar System, except that the motion is not confined to a plane but ranges over all space), then because they are undergoing the centripetal acceleration of that orbital motion, they must radiate electromagnetic energy. A simple calculation showed that they must very quickly collapse onto the nucleus and the Rutherford model would not be stable. This led Bohr to invoke quantum principles to account for the basic structure and stability, and to show that a new quantum mechanics was required for such microscopic scales. Today, we know that quantum mechanics is the governing mechanics of

positively charged nucleus. He, and assistants in his group, developed the first apparatus for accelerating charged particles to high energy to cause nuclear reactions. He also hypothesized the presence of neutral particles in the nucleus, such neutrons being discovered later by his associate James Chadwick.

our Universe, accounting in a manner for everything, whether atoms, mountains, or stars [11], but its effects are especially evident for atoms and nuclei. A planet bound to the Sun can be understood through classical physics, but not the hydrogen atom.

Both physical systems of two bodies, planet–Sun or proton–electron, held together through an attractive $1/r$ potential, gravitational or electromagnetic, respectively, have many things in common. Whether in classical or quantum mechanics, both are well recognized as having more symmetry than expected from the spherical symmetry of the gravitational or Coulomb[22] field. This spherical symmetry is associated with the conservation of angular momentum, $\vec{\ell}$, either expressed as $d\vec{\ell}/dt = 0$, or in quantum mechanics as the angular momentum operator commuting with the energy operator called the Hamiltonian[23], $[H, \vec{\ell}] = 0$. (A commutator of two operations A and B is defined as $[A, B] \equiv AB - BA$ so that its vanishing, or the equivalent statement that the two commute, means that the order in which they are taken does not matter. In quantum physics, only such pairs can be sharply defined simultaneously for any physical system.) Therefore, classical orbits lie in the plane perpendicular to $\vec{\ell}$, as is indeed observed in planetary motion (Figure 1.12). Quantum mechanics does not have orbits and trajectories (dependent in their very concept on both position and velocity, something not allowed by the uncertainty principle because these two quantities do not commute) and, as already noted, the electron's motion should not be pictured in such terms but rather as a probability distribution in all three-dimensional space around the nucleus.

---

[22] Charles Augustin de Coulomb, 1736–1806, French. Retired from the military as an engineer to pursue scientific research and discovered forces between electrically charged objects.

[23] William Rowan Hamilton, 1805–1865, Irish. Mathematician and physicist, and versed in several languages. He made important contributions to optics and mechanics, most notably in reformulating Newtonian mechanics in terms of the energy, or Hamiltonian as now named, and a variational principle. His work on analytical mechanics was uncannily prescient of quantum mechanics, especially in the form it emerged 100 years later in the hands of Dirac. Hamilton's most notable contribution in mathematics was his discovery of quaternions, 'four-dimensional numbers' generalizing two-dimensional complex numbers and based on three square roots of -1. Quaternionic algebra is an alternative to vectors with some advantages in describing rotations and is so used today in orbital mechanics and signal and control theory. Early workers such as Maxwell and some present physicists advocate their use in mechanics and electromagnetism but the dominance of vector mathematics is likely to persist in physics.

Quantum physics, unlike classical physics, also restricts the possible negative-energy bound states of a physical system. Thus, the electron–proton system may have any positive energy upwards from zero and these describe the situation when the electron can separate to infinite distance from the proton, the so-called continuum or scattering (of an electron from a proton) states of the hydrogen atom. Similar continuum states of the classical counterpart, the Solar System, are the parabolic and hyperbolic orbits of a comet or other object that can be flung to infinity from the star, in contrast to the elliptic orbits, wherein the object remains bound.

But, for negative energies, the bound states of the hydrogen atom cannot have arbitrary values as in a classical system. Instead, they are quantized into discrete, allowed energies, whereas any elliptical bound orbit with arbitrary negative energy is a possible state for a planet (of course, quantum mechanics also applies here in principle but, because of the weakness of gravitational interaction relative to the electromagnetic, the spacing between allowed energies is so small as to be negligible and for practical purposes can be treated as continuously distributed). Bohr's elucidation in 1913 of the possible bound-state energies, that agreed with empirical observations made by Balmer[24] based on spectroscopic studies towards the end of the 19th century, was of course one of the first triumphs of quantum physics as applied to matter.

The spherical symmetry of the interaction and that $\vec{\ell}$ commutes with the Hamiltonian are realized through the fact that quantum-mechanical (Bohr) energy levels (also, of course, the continuum energy states) are 'degenerate' in the 'azimuthal' quantum number, m; that is, states of different m share the same energy. This quantum number, m, is a measure of the projection of the angular momentum, $\vec{\ell}$, on the z-axis, ℓ_z, and in a quantum system takes (in units of $\hbar$) integer values between $-\ell$ and ℓ. ℓ itself is also quantized, taking only values of 0 or positive integers. All $(2\ell + 1)$ levels of any ℓ but differing in m have the same energy, which is indicative of the spherical symmetry of the underlying Hamiltonian, that the z-axis is no more distinguished than any other direction. In the language of group theory, which is the mathematics of symmetries, the symmetry is of $O(3)$, the orthogonal (or rotation)

---

[24] Johann Jakob Balmer, 1852–1898, Swiss. A mathematics teacher in a gymnasium, he noticed a pattern in the energies of spectral lines of hydrogen and devised an empirical formula that became a key to Bohr's explanation of atomic structure a quarter century later.

group in three dimensions. Any three-dimensional rotation leaves the $1/r$ potential unchanged because it has no directional aspect, and so is what is termed a scalar. This group is said to have one 'Casimir'[25] invariant, the squared angular momentum with value $\ell(\ell + 1)\hbar^2$. All $(2\ell + 1)$ m-states have this same squared angular momentum.

For bound states, the allowed negative energies are given by the Bohr expression, $-1/n^2$, in units of the Rydberg[26], 13.6 eV (1 eV is the energy gained by an electron accelerated by an electric potential of 1 volt and equals 1.6×10^{-19} J). n is the 'principal' quantum number, taking values from 1 through the positive integers. But the fact that the Kepler[27] orbits for bound states are closed ellipses, and in the quantum treatment that levels of different ℓ but the same principal quantum number n are degenerate (have the same energy), are not explained by the three-dimensional spherical symmetry that only requires degeneracy of m values. For each n, ℓ can range from 0 to $n - 1$. That they are also degenerate points to something additional to rotational symmetry in the Coulomb or gravitational force. A closed ellipse points to the existence of another conserved vector besides $\vec{\ell}$. This vector, $\vec{A}$, points in the direction of the major axis and has magnitude equal to the eccentricity of the ellipse (Figure 1.12). This was already recognized by Laplace[28].

---

[25] Hendrik B. G. Casimir, 1909–2000, Dutch. Theoretical physicist with many contributions to superconductivity, invariants of Lie groups, molecular and nuclear rotations, and quantum zero-point energy forces between objects, both microscopic and macroscopic, some of the latter only recently amenable to experimental measurements. He was a co-founder and director for many years of the Philips research laboratories in his native Netherlands.

[26] Johannes Robert Rydberg, 1854–1919, Swedish. Discoverer of the formula that bears his name for the spectral lines from an atom when it changes from one energy level to another. The fundamental constant of spectroscopy is named the Rydberg. Today, bound states at high energies are named Rydberg atoms.

[27] Johannes Kepler, 1571–1630, German. Inherited Tycho Brahe's observatory and observations of orbital data on planets and found that he had to depart from the circular orbits of the Copernican system to elliptical ones, and formulated the three laws of planetary motion which were the basis for Newton's law of gravitation. He can be credited with having brought mathematical physics into astronomy.

[28] Pierre Simon de Laplace, 1749–1827, French. Mathematician, physicist, and astronomer, who extended Newton's celestial mechanics to consider the stability of and the nebular origin of the Solar System. He formulated Laplace's equation and the differential operator called the Laplacian that occurs in wide areas of classical and quantum physics. He developed potential theory and the 'spherical harmonics' that are used for describing angular dependences in physics and engineering. He invented the Laplace transform, a powerful mathematical technique, and made many contributions

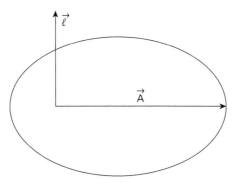

Figure 1.12 A planetary orbit's Kepler ellipse on a plane with a perpendicular angular momentum vector $\vec{\ell}$, and the Laplace–Runge–Lenz vector $\vec{A}$, which points in the direction of the major axis of the ellipse and has magnitude proportional to its eccentricity.

It is the existence of such a conserved vector in the plane that accounts for the orbits closing and thereby the occurrence of a closed elliptical orbit for a planet's motion; otherwise, based only on spherical symmetry and consequent conservation of angular momentum, they could be any orbits so long as they lie in that plane. Indeed, general relativistic corrections that lead to small departures from the Newtonian $1/r$ potential, but still spherically symmetric, retain the planar nature but spoil the closing of the orbit so that the major axis rotates (precesses), albeit very slowly. This was accounted for by Einstein's General Theory of Relativity (Sec. 5.1.2). Similarly, in quantum mechanics, other atoms beyond hydrogen do not exhibit the degeneracy in ℓ, the presence of other electrons leading to small departures from the pure $1/r$ Coulomb field of the nucleus.

To see the nature of this higher symmetry, larger than the obvious isotropic one, of the $1/r$ fields, and that it reflects a symmetry under rotations in one extra dimension as per the theme of this chapter, it is easiest to do so from the Schrödinger[29] equation for the hydrogen atom

to probability and statistics. He also had the first ideas on what were later called black holes, when he argued for massive objects from which even light could not escape.

[29] Erwin Schrödinger, 1887–1961, Austrian. In 1926, formulated the first wave equation of quantum mechanics that bears his name, and then established the equivalence of this wave mechanics to the matrix mechanics of Heisenberg. He went on to develop

written in momentum space. Besides the three dimensions of the momentum vector, $\vec{p}$, the energy as a fourth component can be used to cast the equation as spherically symmetric in all four, that is, as symmetric under rotations in four-dimensional space. The same conclusion follows in coordinate space by the recognition that with both $\vec{\ell}$ and $\vec{A}$ commuting with the Hamiltonian, there are six such operators compatible with a conserved energy. Just as three of the former generate rotations in three dimensions, the six together describe rotations in a four-dimensional space, $d(d-1)/2$ in d dimensions being the number of planes and thereby the number of independent rotations in any dimensional space. The symmetry group is $O(4)$, larger than $O(3)$, which forms a sub-group of it. There are now two conserved Casimir invariants, one expressed by $\vec{\ell} \cdot \vec{A} = 0$, and evidenced by the two vectors lying perpendicular to and in the plane of the orbits (Figure 1.12), and another the extension of the squared angular momentum now to the sum of the squares of both $\vec{\ell}$ and $\vec{A}$. This second invariant is simply related to the energy (or to n for bound states of the atom).

The extra dimension can be geometrically visualized in terms of Hamilton's 'hodograph', the circle that results when the velocity vectors of any Kepler orbit, bound ellipse or open parabola or hyperbola, that is, of all possible positive or negative energies, are plotted together from a common origin (see Figure 1.13). The radius of these circles for the same energy depends on the angular momentum. Collecting these circles provides the extra dimension to the three spatial ones. In quantum physics, this is associated with the degeneracy of different ℓ values at the same energy; note that for bound states, the degeneracy is finite ($\sum_{\ell=0}^{n-1} \sum_{m=-\ell}^{\ell} 1 = n^2$), while it is infinite for continuum states.

Since the components of the two vectors do not themselves mutually commute except for one pair of them, either (ℓ^2, ℓ_z) or (ℓ_z, A_z), one has only a set of three mutually commuting operators along with the Hamiltonian providing the unique labelling of the states of the

techniques for handling perturbing potentials in quantum systems, applying to the effects of electric and magnetic fields on atoms. Widely versed in philosophy, he never accepted the probability interpretation of the quantum wave function. His formulation of 'the Schrödinger cat' (Sec. 4.2.1) as a hypothetical to pose the problems of quantum interpretation continues to capture the imagination of physicist and layman alike. He wrote on consciousness and on the philosophy of biology. His book *What Is Life* on self-replicating systems and on a molecule as the basis of heredity has widely influenced biologists, from the discoverers of DNA structure and the genetic code to today.

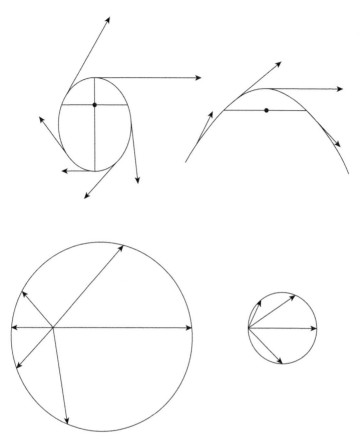

Figure 1.13 The hodograph: circles for elliptical and parabolic orbits (shown on top) when instantaneous velocity vectors along the orbit are gathered together from a common point. That common point lies inside, or on, the circle, respectively, for the two motions with negative and zero energies (the particle in parabolic orbit comes in and recedes to infinity where it has zero speed). A similar construction for hyperbolic orbits of positive energy would have a hodograph with the point outside the circle.

hydrogen atom. There are, however, two equally valid alternative 'representations' (more on this in Chapter 2), depending on the choice between the two sets, called spherical and parabolic, respectively. In either way of counting, the degeneracy of any energy level is larger than the $(2\ell + 1)$ of three-dimensional symmetry and is the well-known n^2 (for bound states) of the higher symmetry. In particular, for $n = 2$, the first excited state of the atom, the four degenerate states group either as (s, p) (the latter embracing three m values) of the spherical representation ($\ell = 0, 1$ are named s and p, respectively, by historical convention) or the two spinors $(\frac{1}{2}, \frac{1}{2})$ of the parabolic, these fractional angular momenta being the quantum numbers of the two mutually commuting operators $\frac{1}{2}(\vec{\ell} \pm \vec{A})$. Recall that $\vec{\ell}$ and $\vec{A}$ are mutually orthogonal, one perpendicular and one parallel to the plane of the or-bit; as a result, these two linear combinations of them are mutually commuting and have equal magnitudes.

   For other discrete energy levels, n, the fractional angular momenta, $(\vec{\ell} \pm \vec{A})/2$, take values $(n-1)/2$, the pair again sharing the same value as a consequence of the orthogonality of $\vec{\ell}$ and $\vec{A}$. Adding the two angular momenta gives all the values $\ell = 0, 1, 2, \ldots n-1$ that occur at that n. The $n = 2$ example illustrates another aspect of adding a dimension, in that, for the three states $2p$, it is upon adding another, the $2s$, that one has the space of four degenerate states that then also splits alterna-tively into the four parabolic states. The next sub-section will provide an analogous example in nuclear spectra.

In coordinate space as well, there is an interesting realization in that the Coulomb problem of the $1/r$ potential in three dimensions can be embedded in four dimensions (u_1, u_2, u_3, u_4) as an isotropic harmonic os-cillator (that is, the pendulum) in those four coordinates as per the transformation [12],

$$
\begin{aligned}
x &= 2(u_1 u_3 + u_2 u_4) \\
y &= 2(u_2 u_3 - u_1 u_4) \\
z &= u_3^2 + u_4^2 - u_1^2 - u_2^2 \\
r &= u_3^2 + u_4^2 + u_1^2 + u_2^2.
\end{aligned}
\tag{1.9}
$$

Each coordinate has a 'spinorial' decomposition as products of two of the u coordinates (which are, therefore, like 'square-root coordin-ates'). In such a rendition of the three-dimensional coordinates with one additional dimension, there is of course redundancy and thereby a constraint, $u_1 u_4 + u_2 u_3 = 0$. For many discussions of the hydro-gen atom, such as its quantum-mechanical path integral treatment or for effects of external electric or magnetic fields on the atom, this four-dimensional system with oscillator-like form proves much more suitable, the motion of a pendulum being much simpler than that in a Coulomb field. This so-called 'regularization' of the Coulomb singularity is already important in classical mechanics and, indeed, the above transformation was actually first introduced by astronomer Kustaanheimo[30] and co-workers for gravitational orbits in celestial mechanics.

The symmetry group of rotations in four dimensions, $O(4)$, with its six generating operators, is the larger (than $O(3)$) symmetry of the discrete Bohr energy states of the hydrogen atom at negative energies -13.6 eV$/n^2$, with a large (again, larger than $(2\ell + 1)$) but finite number, n^2, of states sharing the same energy. (Electronic spin gives an additional doubling and is responsible for the $2n^2$ value that determines the shell structure underlying the Mendeleev[31] Periodic Table of elements.) It is closely related to the symmetry group of Lorentz transformations of space–time discussed in the previous sub-section but with the important difference that the states in that case are infinite in number (see Sec. 1.2.4).

1.2.6 The Interacting Boson Model

The role of an added dimension for a better understanding of atoms in the previous section also has an application in nuclei. A nucleus is a collection of protons and neutrons, fermionic particles. (Because their spin angular momentum is $1/2$, they obey Fermi[32]–Dirac statistics, unlike

---

[30] Paul Kustaanheimo, 1924–1997, Finnish. Astronomer who introduced a method of regularizing the gravitational potential.

[31] Dmitri Ivanovich Mendeleev, 1834–1907, Russian. Chemist who arranged the elements by their atomic weights and produced the Periodic Table, one of the greatest organizing principles of science. He predicted the existence of several 'missing' elements that were later added to their slots in the Table. He made many other contributions to chemistry, investigated the composition of petroleum, although he argued for an abiotic origin from carbon in the deep interior of the Earth, and is said to have helped establish the first petroleum refinery in Russia. He also helped in establishing the metric system in Russia.

[32] Enrico Fermi, 1901–1954, Italian and American. An outstanding physicist, both theoretical and experimental. He established the first theory of beta decay (naming the additional particle emitted the neutrino, which was observed only decades later), a precursor to later interacting field theories. Using neutron bombardment, he created artificial isotopes and elucidated nuclear structure. While missing the first hints, he went on to study fission of uranium, neutron multiplication, and the chain reaction,

particles of integer spin, which obey Bose[33]–Einstein statistics, this being one of the fundamental divisions of our Universe – into fermions and bosons: see Sec. 2.2.1 and 7.3.3.) Instead of the electromagnetic interaction that holds an atom together, 'strong interactions' bind protons and neutrons. This is strong enough to overcome the Coulomb repulsion between the protons in a nucleus.

Bound-state nuclear energy levels are also quantized in a discrete spectrum but are now more complicated than the Bohr spectrum of a simple Coulomb potential. In the low-lying spectral levels of a nucleus and transitions between them, the quadrupole operator plays a major role, just as the dipole operator of electromagnetism does in atoms. The quadrupole corresponds to the angular momentum $\ell = 2$ called d (again for historical reasons, as with s and p) and is viewed as a 'd-boson'. The interacting boson model was advanced as a useful picture of a nucleus, to see it as a collection of bosons rather than as a cluster of the component fermions (combining two fermions gives a bosonic entity). What gave immense power to the model was, however, the adjoining of one more degree of freedom, an s-boson with $\ell = 0$, an added dimension, and to see the low-energy spectral region in terms of such an interacting collection of s and d bosons.

As per the $(2\ell + 1)$ multiplicity, s and d denote one- and five-dimensional objects, respectively. It is this starting picture of six

building the first critical nuclear reactor in the Manhattan Project and going on to the building of the first fission bombs. But, he opposed the further development of hydrogen fusion bombs on both moral and technical grounds. His work on quantum statistical mechanics has led to the naming of all half-odd integer spin particles in the Universe fermions and the statistics they obey as Fermi–Dirac. He also advanced the first models of acceleration of cosmic rays through shocks and varying magnetic fields, still the only viable scenario for the most energetic particles seen today. A gifted expositor, he is known for two influential schools of physicists under him, first in Rome and then in Chicago. His ability to get to the essence of any physics problem, making a first estimate of reasonable accuracy in minutes and on the 'back of an envelope', have led to what physicists refer to as 'Fermi problems'.

[33] Satyendra Nath Bose, 1894–1974, Indian. He re-derived Planck's black-body radiation law in a novel way, entirely within a quantum picture and based on a way of counting identical particles. He sent the paper to Einstein, who recognized its importance, himself translating it and having it published in a German physics journal, and seeing in it a general way of describing identical particles in quantum physics. Applied to photons as the quanta of light and later to all integer-spin particles, they are now referred to as bosons and the statistics as Bose–Einstein.

dimensions and the symmetries associated with them that constitutes the interacting boson model. The group symmetry starts with the unitary group $U(6)$. Since rotations in ordinary three-dimensional space, and the associated group $O(3)$, are obviously involved in labelling the energy states, one looks for 'dynamical symmetry', which means that the energies can be accounted for through the invariant Casimir operators of the relevant groups and sub-groups lying between $U(6)$ and $O(3)$. This is also of great practical significance because calculations reduce to simple algebra upon such restriction to the invariants which do not change in value under various operations.

Between the starting group, $U(6)$, and the final $O(3)$, there are three possible chains of sub-groups and therefore three possible expressions for the energy levels in terms of the Casimir eigenvalues; see Figure 1.14. Remarkably, all three seem to be realized in nature upon examining low-lying spectra of nuclei across the Periodic Table [13]. Thus all such nuclear spectra can be ordered in these three groupings. Even some nuclei that depart from one of the three nevertheless admit simplicities in their description as lying close to one of the three limiting cases and therefore reflecting perturbations about the basic three. This has proved to be an immensely significant organizing principle. Again, for the theme of this chapter, the addition of the s-boson and starting with $U(6)$ rather than $U(5)$ proved crucial, illustrating the importance of the theme of adding a dimension.

More details of this model or of further extensions in boson–fermion models are outside the realm of our discussion but it is interesting to look back at the previous sub-section from this perspective of the interacting boson model. For atoms, the dipole is important, that is, $\ell = 1$ or p in place of nuclei's d, a vector rather than a tensor of rank two. Adding an s to it gives four dimensions, with $O(4)$ symmetry, and the alternative breakups of spherical and parabolic representations, as in the previous sub-section. That same adjoining of a dimension and going to four dimensions is crucial to get the bi-spinor or parabolic decomposition, which would be inaccessible using just the p or three degrees of freedom. The same idea of adding an s but now to a d applies in nuclei with its three primary divisions in Figure 1.14.

Remarkably, the same theme has been extended in atoms in a discussion of the complicated spectra of lanthanides (elements around lanthanum) and actinides (elements around uranium) as pertaining to

$$U(6) \supset U(5) \supset O(5) \supset O(3)$$
$$U(6) \supset SU(3) \supset O(3)$$
$$U(6) \supset O(6) \supset O(5) \supset O(3)$$

Figure 1.14 Three alternative pathways of group $U(6)$ and sub-group chains. Low-lying energy levels of nuclei are well described by the invariant operators of the groups in each chain.

large atoms such as uranium and plutonium. This high up in the Periodic Table, such atoms involve f electrons, that is, $\ell = 3$. Because of the many levels involved with such large angular momenta, and additional complications of fine-structure and other splittings, these spectra are notoriously complicated. Somewhat in the same spirit as the interacting boson model, the suggestion was made to add an s, again adding an extra dimension, to see the $(s + f)$ as an eight-dimensional object and to examine alternative symmetries and simplifications provided by the groups involved [14]. In a further twist, eight is also alternatively viewed as the dimension of the space of three spin-1/2 'pseudo-quarks', each of dimension two, their product eight-dimensional. (This borrowed from 'quarks', invoked as spin-1/2 fundamental constituents of protons and neutrons, a triplet of them forming any nucleon.) Therefore, lanthanide and actinide spectra are described in terms of an underlying three-quark structure, much as $s+p$ was viewed in terms of two spin-1/2. In the language of quantum information (see Sec. 4.2) rather than that of elementary particle physics, wherein spin-1/2 constitutes a 'qubit', we may regard these as three and two qubits respectively. And, in the same vein, the interacting boson model of $s + d$ would correspond to a qubit–qutrit (spin-1) bipartite system with dimension $2 \times 3 = 6$.

1.3 Extra Dimensions to Remove Singularities

Another remarkable use of varying the dimensions involved in a problem is to remove seeming infinities that occur in our mathematical treatment of some problems when it is clear from the physics that there are no such singularities. Quantum field theories that go beyond quantum mechanics in being fully consistent with the Special Theory of Relativity (see Sec. 7.3.3) are notoriously plagued by these infinities involving singular integrals. A simple illustration is provided by the triplet of elementary particles called pi-mesons or pions, a set of three elementary particles comprising a pair that is oppositely charged and a neutral one. An immediate expectation is that the small difference in mass of the neutral particle from that of the charged pair (which have equal masses, an aspect of what is called CPT symmetry and required of all particle–anti-particle pairs; see Sec. 5.2.1) is due to the electromagnetic interaction in which they differ. Indeed, the difference in mass of about 1% connects plausibly with the expectation based on the relative weakness of electromagnetism compared with strong interactions, which are otherwise dominant and thereby contribute most to the mass. However, a first calculation of the difference comes out infinite. Various 'renormalization' methods are now familiar to overcome such singularities.

One class of them is termed dimensional renormalization, wherein the problem is treated with extra dimensions [15]. Generally, these discussions are handled in momentum space but to convey the essence in simpler terms, consider the integral of $1/r$ in ordinary calculus. Although this function is singular at the origin, if r stands as usual for distance, in three dimensions there is no infinity from the short-distance behaviour because of the powers of r in the volume element $r^2 dr$ of integration over three dimensions. On the other hand, in one dimension, $\int_0 dx/|x|$ would indeed be infinite, a divergent integral. Most divergences in field theories arise indeed from high-momentum or, equivalently, short-distance behaviour (the link between small distance and large momentum is an essential feature of quantum physics) so that these simple observations are indeed relevant. With that, we turn to a concrete problem in quantum mechanics based on the above $1/r$ observation to illustrate the point of dimensional renormalization.

The hydrogen atom is a system with a Coulomb potential, $-e^2/r$. Quantum mechanically, it is stable because the quantum kinetic energy prevents the electron from getting arbitrarily close to the nucleus, that is, $r = 0$. Instead, on average, the electron is held to a distance of about the Bohr radius, $a_0 = \hbar^2/me^2 \approx 5 \times 10^{-11}$ m, sees only that much of the attraction of the nucleus, and ends up with a corresponding binding energy of -13.6 eV in the ground state. Consider, however, an atom in very strong magnetic fields, such as have been found on neutron stars and magnetars. These fields are not only larger than on any other objects in our Universe but they are overwhelmingly stronger than the internal electric and magnetic fields in the atom. As a result, such a field completely changes atomic structure. It confines the electron's motion in the two directions perpendicular to the magnetic field, that field controlling those motions and dwarfing the Coulomb force from the nucleus in those directions. On the other hand, because a magnetic field, no matter how strong, exerts no force in the direction parallel to itself, the Coulomb attraction by the nucleus operates in that direction to bind the electron. Effectively, the Coulomb binding operates only in that one dimension [16].

Because the electron by virtue of being confined in the other dimensions sees much more of the nuclear field, its binding is indeed enhanced. If it were truly one dimensional, by our above argument of the divergence of a one-dimensional integral, we would conclude that there is infinite binding or collapse of the electron onto the nucleus. However, the actual problem is in three dimensions and calculations show enhanced but not infinite binding. Further, the enhancement is logarithmic, the argument of the logarithm being the ratio of the Bohr radius to the cyclotron radius. This can be understood in the language of dimensional renormalization by arguing that, while appearing mostly one dimensional, at very short distances (within the magnetic cyclotron radius, which, while smaller than a_0, is still of some non-zero value) the r^2 of the three-dimensional volume element prevents the divergence. Here, dimensional renormalization of a 'one-dimensional hydrogen atom' is not a mathematical device but in the very physics of the structure of atoms on neutron stars.

Turning the above argument around, there are speculations in physics about whether our own world of three space dimensions may actually have extensions into extra dimensions, typically considered tiny, as in the above example [7]. Thus, every point of our three-dimensional world may actually be a very small rolled-up circle into another dimension (similar to the hodograph of the hydrogen atom in a previous sub-section where they were not a real new space dimension). Such speculations have been invoked for a variety of consequences, renormalization being just one. Another, for instance, is that the seeming weakness of some interaction (such as gravitation compared with the rest) may be only a reflection of the different amount by which some may act in the three space dimensions we see while spilling over mostly into unobserved dimensions.

Yet another is to explain why some of the fundamental constants that characterize our world have the values they do. Thus, the enhanced binding, larger than 13.6 eV, of the hydrogen atom in a strong magnetic field may be interpreted in terms of a stronger effective e^2, being a combination of the usual charge of an electron and the strength of the magnetic field that also enters into the expression for the binding energy. Conversely, what appear in our world as fundamental constants, such as the charge or mass of the electron, may actually reflect some fields in the larger dimensions that have 'compactified' all but the three we observe, much as the magnetic field confines from three into the one dimension along its direction for atoms on neutron stars. The values we observe may thus contain the particular value (of no special significance) of those fields and there may, therefore, similarly be no special significance to them [16].

2

Physics as Transformations

2.1 Introduction to Transformations

Transformations, from one coordinate system to another, from one observer's frame of reference to another, from a description in one basis to another, are all inherent to physics. Indeed, even outside of physics, there are famous and basic roles for transformations, as in Felix Klein's[1] 'Erlangen Programme', which completely re-oriented the study of geometry. Instead of the Euclidean approach, familiar from high school, of specifying axioms, followed by theorems about points, lines, triangles, circles, and other figures, Klein emphasized that one defines a set of symmetries and transformations and for each such set there exists a geometry. Euclidean[2] geometry that results from Euclidean transformations in a plane is but one of many geometries. A different set of symmetry transformations will define a different geometry. The intimate connection between symmetries and transformations is reflected also in the material on symmetries in Chapter 5, and this chapter overlaps with that material and with that on maps in Chapter 6.

Mechanics, the very first subject in physics, deals with the motion of a physical system, through either translation or rotation, and all of mechanics can itself be viewed in terms of transformations under such operations, the system changing from an initial state to a final one. This

---

[1] Felix Klein, 1849–1925, German. Mathematician known for his work in group theory, analysis, and geometry. His Erlangen Programme that he laid out in his inaugural lecture at Erlangen revolutionized the study of geometry. He established one of the great schools of mathematics at Göttingen and launched the *Encyclopaedia of Mathematics*. He wrote a well-known book on the icosahedron, and a topological object has been named the 'Klein bottle'.

[2] Euclid of Alexandria, circa 300 BC, Greek. A mathematician who built on works before him to give a coherent presentation of plane geometry through a few axioms and rigorous proofs. His principal work, *Elements*, also has results in number theory and has had a central role in mathematics for centuries.

point of view makes transformation theory already a central feature of classical mechanics but it becomes even more of an immediate starting point in quantum mechanics. Before considering various aspects of transformations in quantum physics, it is worth noting an even older philosophical recognition in every culture and civilization that transformations dominate nature and human life.

Ovid's[3] *Metamorphoses*, systems of yoga in India, ancient myths of many religions and cultures, and Grimms'[4] *Fairy Tales* are all full of transformations of man, beasts, and gods. A striking and charming expression is Chuang Tzu's[5] butterfly: 'Once I, Chuang Tzu, dreamed I was a butterfly and was happy as a butterfly. I was conscious that I was quite pleased with myself, but I did not know that I was Tzu. Suddenly, I awoke, and there was I, visibly Tzu. I do not know whether it was Tzu dreaming that he was a butterfly or the butterfly dreaming that he was Tzu. Between Tzu and the butterfly there must be some distinction. This is called the transformation of things.' [17] Other cultures and their mythology also play with transformations, exploring very pliable changes in shape, size, time, gender, number, and other familiars, under some astonishingly imaginative transformations.

The simple example (Figure 1.5) at the beginning of Chapter 1, of equivalence between simple harmonic motion in a line and circular motion on the circle with that line as diameter, provides a good illustration of the advantage of transforming between different coordinate systems. Instead of the coordinate x along the horizontal, and its companion coordinate y along the vertical, these 'Cartesian'[6] coordinates may be replaced by the 'circular coordinates' (ρ, ϕ) of that

[3] Publius Ovidius Naso, 43 BC–18 AD, Italian. Great Latin poet, prolific writer on Roman culture, politics, and religion. His love poems and text on transformations in Greek and Roman myths are especially well known. One of his tragedies, *Medea*, has inspired numerous stage and opera productions to this day.

[4] Jacob Grimm, 1785–1863, and Wilhelm Grimm, 1786–1859, German. The brothers were linguists and collectors of folklore, and started publishing German folk stories from 1812. These have become among the best-known folk stories across much of the world, mined also for their moral and psychological truths. They also started a major German dictionary project.

[5] Chuang Tzu or Zhuang Zhou or Zhuangzi or Master Zhuang, circa 4th century BC, Chinese. A Daoist philosopher who wrote the book *Zhuangzi*, a philosophy of scepticism.

[6] Rene Descartes, 1596–1650, French. Philosopher and mathematician, and considered one of the fathers of philosophy. He is also considered the originator of

two-dimensional world through

$$x = \rho \cos \phi, y = \rho \sin \phi, \qquad (2.1)$$

and, equivalently, the inverse transformation,

$$\rho = \sqrt{x^2 + y^2}, \phi = \arctan(y/x). \qquad (2.2)$$

Instead of a (Cartesian) grid of horizontal and vertical lines parallel to the x and y axes, the plane is covered by concentric circles of different radius around the origin and radiating straight lines from that origin. These two sets, circles and radial lines, are also mutually perpendicular, 'orthogonal', wherever they meet. Both rectangular and circular coordinates cover the plane and are equally capable of describing it (see Figure 2.1).

This is a simple instance of later examples in this chapter that alternative representations can in their entirety cover the system under consideration, here the two-dimensional space of the plane. D'Arcy Thompson, who was mentioned in Chapter 1, gives more complicated grid transformations that put skulls such as of a chimpanzee and a human on an equivalent footing. In today's age of computers, such 'morphings' with even more complexity appear quite commonly on our TV screens. The first creature to crawl out on land morphs in a few frames to cover intermediate steps towards a walking human in a cartoonish stand-in for Darwin's[7] biological theory of evolution!

Simple harmonic motions in either x or y with amplitude A and frequency ω are described by (an over-dot will denote differentiation with respect to time)

$$\ddot{x} = -\omega^2 x, \qquad (2.3)$$

with $\omega = \sqrt{g/\ell}$ for the pendulum in Figure 1.3. Substituting Eq. (2.1) in Eq. (2.3) and equating terms in sin and cos on either side gives

analytical geometry, which combines algebra and geometry, and the system of coordinates for a point in two- or higher-dimensional space is named for him. He was a major figure in 17th-century rationalism. He had many works in mathematics and philosophy. His statement *Cogito er-go sum*, or 'I think, therefore, I am', is often quoted.

[7] Charles Darwin, 1809–1882, English. Naturalist and biologist, discoverer of natural selection as the mechanism for the evolution of different biological species. It is considered the unifying theory of the life sciences. His 1859 book *On the Origin of Species* and other books and insights stand at the centre of modern biology.

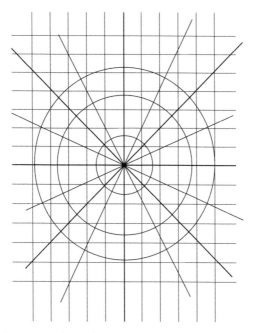

Figure 2.1 Cartesian and circular grids to cover the x–y plane. In the former, parallel horizontal and vertical lines form a mutually perpendicular set to assign coordinates to any point; in the latter, concentric circles and radial lines do so, again intersecting at right angles. Also shown are 45-degree axes besides the Cartesian (x, y) axes.

$$\ddot{\rho} - \rho(\dot{\phi})^2 = -\omega^2 \rho, \tag{2.4}$$
$$\rho\ddot{\phi} + 2\dot{\rho}\dot{\phi} = 0.$$

While these equations may at first sight look more complicated than the Cartesian Eq. (2.3), they prove especially convenient in many situations and not just in the example of Chapter 1 with fixed frequency $\omega = \dot{\phi}$ and amplitude A. In that case, with constant values for $\dot{\phi}$ and ρ, and all further dots (time derivatives) on them zero, Eq. (2.4) is trivially satisfied, requiring no further solution.

But, quite generally, through simple steps of differential calculus, the second of the equations in Eq. (2.4) reduces to vanishing of the differential (single dot) of

$$\rho^2 \dot{\phi}. \tag{2.5}$$

This vanishing of the time derivative of the above quantity means that the quantity is constant in time, or 'conserved'. Such conserved quantities or, equivalently, conservation laws, are among the most fundamental of concepts in physics. The quantity in Eq. (2.5), when multiplied by the mass of a particle that may be executing such motion, is called its angular momentum; note its dimensions $[M][L]^2[T]^{-1}$. The law of conservation of angular momentum is one of the absolutely valid conservation laws of physics, together with the laws of conservation of linear momentum and energy.

The conservation of angular momentum in the case of the example considered in Chapter 1, when both $\dot{\phi} = \omega$ and $\rho = A$ are fixed, is of course trivial, their combination in Eq. (2.5) also necessarily constant. But the above derivation holds more generally. So long as the equations in Eq. (2.4) do not involve ϕ so that, even in complicated situations of other forces acting on the pendulum (or two-dimensional oscillator), as long as they depend only on ρ, the conclusion holds that the angular momentum in Eq. (2.5) is conserved. Individually, ρ and ϕ may not but their combination in Eq. (2.5) is constant in time during the motion. This was one of the fundamental insights of Newton, that Kepler's elliptic orbits and, in particular, that equal areas are swept out in equal time so that the planet speeds up when close and slows down when far from the Sun in such a way as to keep Eq. (2.5) constant, revealed something fundamental about the force law governing the motion, that it had to be a 'central force', acting purely radially and independent of angular position in orbit.

Such forces or potentials in physical systems are called isotropic, that is, they are independent of angular direction, and depend only on the radial variable. The association of the conservation of angular momentum with such isotropy is among the most important theorems of physics (see Sec. 5.1.2). It holds not just in the two-dimensional example above but also in higher dimensions. In particular, in our world of three space dimensions, such potentials are called spherically symmetric, a sphere being such an object with no direction singled out as special. Forces and potentials of a spherically symmetric system are independent of directions in space, and depend only on the separation distance involved. Under such forces, the angular momentum of a system remains unchanged, is an invariant of the motion.

Even for time-dependent frequencies, when neither $\dot{\phi} = \omega$ is constant nor ρ, their combination in Eq. (2.5) is. Upon multiplying by mass and denoting it by its usual symbol for angular momentum, ℓ, the first of the equations in Eq. (2.4) can be written as an equation involving ρ alone,

$$\ddot{\rho} - \ell^2/m^2\rho^3 = -\omega^2\rho. \tag{2.6}$$

Again, for constant ω and amplitude, this is trivial, the first term vanishing, but its importance lies in the more general situation of complicated dependences on time of frequency and amplitude. Nevertheless, the combined quantity ℓ is conserved.

Further interest lies in the fact that the passage from Eq. (2.3) to Eq. (2.4) is valid for any linear second-order differential equation. In particular, it is so for the time-independent quantum Schrödinger equation in place of Eq. (2.3). The derivatives are now with respect to coordinate space, not time, and the dependent variable is not the position, x, but the wave function, ψ. But upon the same re-writing of ψ as now the amplitude and phase of the wave function, the subsequent derivation carries through. These counterparts of Eq. (2.4), called the 'phase-amplitude' equations, have proved useful in quantum physics (Sec. 7.4). Part of the reason for this is that, unlike the wave function, ψ, the amplitude and phase are directly accessible to our measuring apparatus. This is an instance of how even when equivalent, transforming from one description, in terms of (sometimes, even complex) wave functions, to another with real quantities of amplitude and phase may have merit, sometimes also philosophical, in being more directly connected to the observables that physics deals with.

2.2 Alternative Representations in Quantum Physics and Transformations Between Them

Both in geometry and other mathematics and in classical physics, alternative representations and transformations between them have been studied from the very beginning. A principal aim of physics is to trace the evolution of a system from its state at some initial instant to a later one under the action of specified forces. The state of a physical system in classical mechanics is specified by giving the positions and velocities of all the masses constituting the system, whether in the Newtonian description or in its later reformulation by Lagrange in terms of a stationary principle for the Lagrangian (Sec. 1.2.3). Instead of talking of forces, which are vector quantities, the reformulation deals only with scalar energies, the Lagrangian being usually the difference between kinetic ($mv^2/2$ or $p^2/2m$) and potential (such as mgh for gravitation or $kx^2/2$ for a spring) energies of the system. Instead of Newton's equations, the equations of motion called Euler[8]–Lagrange equations are first-order differential equations in time along with first-order partial differentials of the Lagrangian with respect to coordinates and velocities. Regardless of the choice of coordinates, these equations always have the same form, which is not so with Newton's equations. For these various reasons, the Lagrangian formulation is more powerful and useful than the Newtonian, although they are equivalent.

Given this 'form invariance', the Euler–Lagrange equations permit easier transformation than do Newton's laws of motion from one set of coordinates to another, such as from Cartesian to circular, or their counterpart spherical in three dimensions. A second, slightly different but closely related formulation in terms of Hamiltonians and Hamilton's equations of motion replaces the Lagrangian velocities by their equivalent momenta and the Lagrangian by the Hamiltonian, which is defined as a function of coordinates, momenta, and time. In most cases, the Hamiltonian is the sum of kinetic and potential energies. Again, transformations allow for a wide variety of coordinates and momenta.

The further step into quantum mechanics makes transformation theory a decisive element, even more than in classical mechanics. This is because of a central characteristic of quantum physics: that both coordinates and velocities (or momenta) cannot be simultaneously specified. This restriction, imposed by the Heisenberg uncertainty principle, arises of course from the nature of our world, that it has a non-zero value of Planck's quantum constant, $\hbar$, a quantity with dimensions of position multiplying momentum. It forces the physicist to make a choice, in even the simplest of physical systems considered, to use either positions or momenta (or some other quantity) in terms of which to view the system. Indeed, quantum physics de-emphasizes which of these

---

[8] Leonhard Euler, 1707–1783, Swiss and German. One of the greatest and most prodigious mathematicians of all time, with wide contributions in many areas. Laplace is said to have expressed Euler's influence on mathematics by saying that 'he is the master of us all'. He also invented much of the notation and terminology, two mathematical constants, e and γ, and the symbol i for the square root of -1, all of which, along with many of his other results and theorems, occur widely throughout mathematics and physics.

quantities is chosen, such 'representations' being matters of choice and convenience but any one of them being equally valid to capture the physics of the system and its dynamics. The question also arises naturally of how to pass, transform, from one description or representation to another.

The fundamental difference between classical and quantum physics lies in what is meant by the physical system. The quantum system is not specified by its coordinates and momenta, which are real measurable quantities, but by a wave function that is complex and itself, therefore, inaccessible to our measuring apparatus. Actually, in an even further step, states may be viewed as abstract vectors in what is called a Hilbert[9] space, independent of any particular wave function description. It remains true that this is a space of complex, not real and measurable, entities. Only bilinear combinations of wave functions and their complex conjugates provide the observed real quantities. The wave function may be expressed in terms of one language or another, position or momentum representations, or perhaps by even other choices such as matrix and other representations.

Thus, considering the simplest example of a one-dimensional harmonic oscillator, the pendulum of Sec. 1.1, its quantum-mechanical wave function may be taken to be functions of x (products of Gaussian functions and another standard set of functions called Hermite[10] polynomials), or functions of momentum, p (in this symmetric problem where both coordinates and momenta enter quadratically in the harmonic oscillator's energy or Hamiltonian, $H = p^2/2m + m\omega^2 x^2/2$, they are also Gaussians and Hermite polynomials, but now in p), or as

---

[9] David Hilbert, 1862–1943, German. Mathematician, one of the greatest and most influential of the 19th and early 20th century. Contributed to many areas of mathematics, including the theory of invariants, set theory, transfinite numbers, functional analysis, and axiomatic geometry. Established a famous mathematical school at Göttingen, editing the major mathematical journal of the time, and collaborating with fellow physicists who were developing the new quantum physics. He developed rigorous mathematical tools for physics, the then newly published text *Courant-Hilbert* readily providing the needed mathematics as quantum mechanics was developed in the 1920s and 1930s. Known as the founder of proof theory, he drew up a list of 23 outstanding mathematical problems that has set the course of subsequent mathematical research, many prominent mathematicians having tackled these 'Hilbert problems'.

[10] Charles Hermite, 1822-1901, French. Mathematician with contributions to number theory, orthogonal polynomials, and numerical analysis. Established the transcendental nature of Euler's number e.

infinite-dimensional matrices, or, even more abstractly, to be discussed below, in terms of Dirac's 'bras' and 'kets'. All physical observables, whether position, momentum, energy, or any combinations of them, act on these wave functions to provide other functions, and the Born interpretation gives the probabilities of specific values that may be observed upon any measurement (Sec. 1.2.2).

Given any one representation, all the physics of the system is contained in it and one can choose to work within that representation alone, as stated earlier about Cartesian or circular coordinates. But, for convenience or other reason, if one wishes to pass to another representation, a transformation can be specified. This is much like different languages, with dictionaries allowing us to go from one to another. The English 'translation' and German 'übersetzung' have precisely this meaning. The probability interpretation of the wave function (Sec. 1.2.2) views its squared norm, $|\psi|^2$ (product of ψ and its complex conjugate), as the probability of obtaining that value, whether of position or momentum or some other quantity, depending on the representation. The requirement of preserving normalization, that is, the squared norm of the wave function integrated over all of the corresponding space must equal unity, makes such transformations 'unitary'. Going from coordinate to momentum representation, or vice versa, is thus given by one such unitary transformation that has long been known in mathematics and physics, the Fourier[11] transformation, x and p being called canonical conjugates. This important transformation between a pair of conjugate variables associates for each function in one variable a corresponding function in the other. The differential operation in one becomes a simpler algebraic multiplication by the variable in the conjugate space. Together with other such useful properties, the Fourier transformation is one of the most important mathematical techniques in physics and engineering.

The conjugate pair of x and p, with commutator $[x, p] \equiv xp - px = i\hbar$, with $\hbar$ Planck's constant, expresses that each member of the pair acts as

---

[11] Jean Baptiste Joseph Fourier, 1768–1830, French. Mathematician and physicist who discovered one of the most important techniques for physics and engineering during his work on heat transfer. He was part of Napoleon Bonaparte's expeditions to Egypt and made governor of Lower Egypt, and was influential in Champollion's translation of the Rosetta Stone. Fourier also contributed to dimensional analysis and was an early discoverer of the greenhouse effect, recognizing that the Earth's surface would be much cooler were it not for its atmosphere.

the derivative or gradient in the other representation. This is why the order in which they act matters and the commutator, defined as the difference between xp and px, is non-vanishing. In the latter, the differential acts also on the factor x that follows it, giving an additional contribution. The difference involves the imaginary unit and Planck's quantum constant, both signifiers of quantum physics. It is a matter of empirical observation that our Universe has a non-zero value of $\hbar$, a quantity with dimensions of position×momentum, that is, of angular momentum $[M][L]^2[T]^{-1}$. This is what makes ours a quantum world. Had $\hbar$ been zero, ours would have been a classical world. However, because of its small value (on the scale of most angular momenta encountered in everyday experience) quantum effects were not appreciated till just over 100 years ago.

Indeed, had physicists encountered quantum physics first, position and momentum might never have been defined as independent entities! Rather, momentum may have simply been seen from the start as the gradient (derivative) in position space, $\vec{p} = (\hbar/i)\nabla$. The existence of the dimensional physical constant $\hbar$ is crucial for permitting such an association between p and x, giving physical context and significance to what were already recognized as a Fourier conjugate pair. The uncertainty principle is, of course, another direct consequence of the non-zero value of $\hbar$ and of the basic commutator between p and x being proportional to it. Actually, as well known with Fourier conjugates, whether x and p, or time, t, and frequency, ω, tight concentration in one translates into a wide distribution in the conjugate variable/space. Note that the dimensions of $\hbar$ may also be thought of as energy×time, and this becomes relevant in Chapter 7's discussion about the nature of time.

This Fourier connection was a precursor in classical physics or mathematical analysis to the uncertainty principle, the 'only' extra (but, of course, crucial) ingredient brought in by quantum physics being (besides i) the non-zero $\hbar$ that connects dimensionally the conjugate quantities. Yet another pair of such conjugate entities are angle and angular momentum, their product also having the dimensions of $\hbar$. Again, tight angular beaming, as from an antenna, is achieved only by superposing many 'multipoles' of angular momentum, ℓ, whereas a single value of ℓ does not pick out unique directions. In particular, $\ell = 0$ describes an isotropic distribution with no directional dependence and no direction singled out, being all on an equal footing.

Quantum physics similarly views angular momentum as a gradient in angles, again with multiplicative $\hbar/i$. It is an important point, born out of such conjugate pairs, that to define any distribution sharply in angles requires superposition of many and large values of ℓ. A spectacular realization of this in recent times is the so-called WMAP (Wilkinson[12] Microwave Anisotropy Probe) observation of small-scale anisotropies in the cosmic microwave background, at the level of 1 in 10^4, analysed in terms of ℓ into the thousands, that reflect primordial fluctuations of very small angular scales in the early Universe which later lead to the large-scale structures, including clusters of galaxies, that we see today.

Actually, perhaps, it might have been more appropriate, given the fundamental law of conservation of momentum and that there is no such for position, to have seen momentum as the basic object. Position could then have been viewed as simply the gradient with respect to it, with a multiplicative $i\hbar$. In non-relativistic quantum physics, and even in relativistic quantum mechanics as in the Dirac equation for an electron or proton (Sec. 7.3.2), both position and momentum are operators, as is energy, while time is not. In quantum field theories, both position and time are seen as merely parameters, thus placing them on an equivalent relativistic footing (relativistic quantum mechanics of a particle is internally inconsistent). We will return to this in Chapter 7 (Sec. 7.3.3).

Since linear momentum and energy are operators in quantum physics, it might have been appropriate to view p and E as the basic elements, and x and t then as derivatives with respect to them (with appropriate factors of $\hbar$ for dimensional reasons and the imaginary element to preserve the so-called Hermitian nature of these operators that is needed to give real values for energy and other physically measurable quantities). Indeed, in a time-independent approach to quantum scattering theory with stationary states of energy, E, and invoking no complex quantities but using only real, standing waves, the 'Wigner[13] time delay'

---

[12] David Todd Wilkinson, 1935–2002, American. Astronomer and cosmologist.

[13] Eugene Paul Wigner, 1902–1995, Hungarian and American. Theoretical physicist with decisive contributions to nuclear and elementary particle physics, especially to the role of symmetry in quantum physics. He introduced and developed the use of group theory, especially for angular momentum in quantum mechanics, and for spin and isotopic spin in nuclear structure. He was a member of the team that developed the first chain reaction in uranium fission and went on to become the physics consultant for the first commercial nuclear power reactors. He was also interested in philosophical

is defined as $2\hbar d\delta/dE$, involving a derivative with respect to energy of δ, the scattering phase shift. We will return to this for another theme in Chapter 7 (see especially Sec. 7.3.1).

2.2.1 Forty-Five-Degree Rotation, Hyperspherical Coordinates, and Correlations

Beginning students of physics learn a standard 'trick' in handling a harmonic oscillator in two dimensions or two coupled one-dimensional oscillators. When decoupled, the system Hamiltonian, $H = (p_x^2 + p_y^2)/2m + (m\omega^2/2)(x^2 + y^2)$, splits of course into two independent, identical pieces, that is, motions in x and y, that are trivially solved separately and then combined into the full solution. Adding a coupling term, kxy, no longer permits such simple 'separation' of variables, but the trick lies in recognizing that changing coordinates to $(x + y)$ and $(x - y)$ recasts the Hamiltonian as two separate oscillators in these '45-degree' coordinates. They are obtained by rotating the Cartesian pair (x, y) through 45 degrees in that plane (see Figure 2.1). A classical pendulum that can swing in the two dimensions can also oscillate about these 45-degree lines, with $y = \pm x$, the two oscillations 'in phase' or exactly 'out of phase'. Even further, for other 'phase angles' between them, the bob executes the motion of a 'conical' pendulum. This is the situation in Chapter 1's Figure 1.5 when the two motions in x and y are exactly out of phase and that circle describes the motion of such a circular or conical pendulum's bob.

This simple 45-degree transformation to rotated axes appears widespread in mathematics and physics. Especially in quantum physics, where special significance attaches to the identity of particles such as electrons, protons, or photons, this transformation takes on further importance. Consider two-electron physics in an atom or two-nucleon phenomena in nuclei. Their mutual interaction renders the system Hamiltonian non-separable in independent coordinates $\vec{r}_1$ and $\vec{r}_2$, or in the so-called independent-particle representation. Thereby, the independent-particle operators such as the particles' individual angular momenta do not commute with the Hamiltonian and cannot be

questions about quantum physics and consciousness, introducing 'Wigner's friend' as a variant to the 'Schrödinger cat' discussion (Sec. 4.2.1) of quantum interpretation. His essay 'The unreasonable effectiveness of mathematics in the natural sciences' has become a classic among mathematicians and physicists.

ascribed definite values or quantum numbers simultaneously with the energy of the system. Only the overall spin, $\vec{S}$, and orbital angular momentum, $\vec{L}$, of the whole system commute with the isotropic H and have meaning, states of the system labelled $^{2S+1}L_J$. When spin–orbit coupling is strong, even $\vec{S}$ and $\vec{L}$ are not individually conserved, only the total angular momentum, $\vec{J} = \vec{S} + \vec{L}$, and quantum numbers of J^2 and J_z label the states along with their energy.

For low-lying states in a many-electron atom, however, with the inter-particle interaction perturbatively small relative to the central field that each sees (in an atom, that of the positively charged nuclear core), the independent-particle labels are a fairly good description and one often uses their 'configuration' labels for the state, as in $1s^2\,^1S_0$ or $1s2s\,^{1,3}S_J$ for the ground and first (singly) excited states of the helium atom. The individual electron labels, in lower case, follow those of the hydrogen atom (see Sec. 1.2.5), with $n = 1, 2, \ldots,$ and $\ell = 0, 1, 2, \ldots$ referred to, respectively, as $s, p, d, \ldots$.

In this set of six coordinates, when viewed as two radial distances and four angles, three of the angles are of less dynamical significance to the physics of the two-particle system than the fourth, namely, the angle between the two radial vectors, called θ_{12}, which determines the separation between them and thereby their interaction. The set of the other three angles, typically chosen as 'Euler angles', defines the orientation of the triangle formed by the two particles and fixed centre of mass of the system in some space-fixed set of three orthogonal axes. It is an element common to all two-particle systems, regardless of their specific dynamics.

The interaction between the two particles, and therefore the critical part of the wave function, depends on r_1, r_2, and θ_{12}, which may be regarded, therefore, as the 'dynamical' variables of the system. They define the triangle regardless of its orientation in space. The wave function is not generally separable as a product of wave functions of the individual particles, so that such a product does not describe a physical state. However, such products in the independent-particle representation for each configuration provide a complete basis set in terms of which to describe the physics of two electrons. Whereas a single configuration, or superposition of a handful, may suffice for low-lying states, in general it requires a superposition of many such basis functions to give a good description for higher states. This is regarded as expressing correlations between the particles.

To handle correlations, we adapt the idea of circular coordinates used above for two different space dimensions to the case of coordinates of two particles. Consider instead of r_1 and r_2 the pair of coordinates,

$$R = (r_1^2 + r_2^2)^{1/2}, \quad \alpha = \arctan(r_2/r_1), \tag{2.7}$$

or, equivalently,

$$r_1 = R\cos\alpha, \quad r_2 = R\sin\alpha. \tag{2.8}$$

Clearly, as in Eq. (2.1) and Eq. (2.2), these are circular coordinates instead of Cartesian in the plane (r_1, r_2). The one difference from Figure 2.1 is that now the two coordinate distances, themselves three-dimensional distances and not one-dimensional coordinates, do not become negative and only the positive quadrant of Figure 2.1 applies.

For the two-electron system, the alternatives of independent-particle or pair coordinates, and of complete sets of basis state functions in either, are two alternative representations, just as are the coordinate and momentum space representations of a single particle. In terms of the pair coordinates in Eq. (2.7), the consequences of quantum physics for identical particles can be handled easily. Among these consequences is that, unlike in classical physics, mere labelling of the particles as 1 and 2 has no quantum meaning. Meaning is attached only when some physical observable can be associated with the labels we give to a physical system, corresponding quantum numbers then identifying the state of the system. Indeed, a complete set of quantum numbers arising from a complete set of operators that mutually commute with each other, when placed as labels inside a Dirac bra or ket denoting the system (see Sec. 2.3), identifies fully a quantum-mechanical state.

A further element is a powerful axiom of quantum physics, called the Pauli[14] Principle, that in a many identical-particle system, interchanging

---

[14] Wolfgang Pauli, 1900–1958, Austrian and Swiss. A crucial figure in the development of quantum theory through especially his interactions with Bohr, Dirac, and Heisenberg. Known for his strong views, and caustically expressed (a Pauli-ism is to dismiss something as 'not even wrong'), he had a very critical sense (the 'conscience of physics') and, already as a young student, wrote a book on Einstein's General Theory of Relativity. He introduced a crucial new quantum number in understanding atomic spectra, later identified with quantum spin, and the associated 'Exclusion Principle'. It forms the basis of all atomic structure and chemistry, and the stability of matter. His connection of spin and statistics is a fundamental element of all quantum field theories. He also developed the regularizations or renormalizations necessary to eliminate

coordinate labels for any two identical particles in the wave function must satisfy one of two possibilities. If the particles are fermions, the name given to those with half-odd integer spin, as in the case of electrons or nucleons, then the wave function must be antisymmetric (change sign) under such interchange, whereas for bosons, the name for particles with integer spin, the wave function must be symmetric. It is this 'spin-statistics theorem' and consideration of interchange that makes the coordinates (R, α) particularly suitable. The Pauli Principle is an aspect of the physics of the world around us, as fundamental as any of the laws of conservation, such as of charge, energy, momentum, etc.

For handling symmetry/antisymmetry, particle interchange leaves R unchanged while $\alpha \rightarrow \frac{\pi}{4} - \alpha$. In particular, the 45-degree line in that plane serves as a symmetry line for radial interchange. Radial wave functions may be either symmetric or antisymmetric under this interchange, which is expressed as reflection in the line (angular and spin interchanges should then behave appropriately so as to ensure overall symmetry or antisymmetry for the spin-statistics requirement), which means the radial functions have either a node (vanish) or antinode (the derivative vanishes) on that line. This means that the probability of finding the particles together is diminished or enhanced, respectively. This has important consequences. If there is a repulsive interaction between particles 1 and 2, an antinode costs energy whereas a node reduces it; the opposite applies for attractive interactions. Thus, the symmetry principle for identical quantum particles can amount to an effective force between them.

Such effects are manifest in both atoms and nuclei. In an atom such as helium with two electrons, for states with both having orbital angular momentum zero, we need only consider the radial and spin wave functions. If the spin angular momenta of the two electrons are coupled

infinities in these theories. 'Pauli spinors' and 'Pauli matrices' provide the central notation and language of quantum-mechanical spin. To explain beta radioactivity and the seeming non-conservation of energy and angular momentum, he postulated a third particle emitted in that beta-decay, later named the neutrino. Neutrino physics is a major part of elementary particle physics to this day. Besides his contributions to philosophical elements of quantum physics, he was seriously interested in psychoanalysis and had a close association with his neighbour, Carl Jung, and their extensive letters have been published.

into a 'singlet' (total spin zero, there being only one such state), the spin wave function is antisymmetric under interchange (see Sec. 4.2.4 and Figure 4.9). Therefore, the radial part has to be symmetric with an anti-node along the 45-degree line which means a higher probability of the two electrons being on top of each other. This raises the energy relative to similar triplet states (total spin one with three possible states) that have a node and thereby a decreased electron–electron repulsion. This 'exchange' splitting puts triplets about 1 eV lower in energy than singlets among the low-lying singly excited states of helium (contrast with the first electronic excitation in helium of about 20 eV). See Figure 2.2.

Note that the exchange interaction arises actually from the electrostatic repulsion among electrons but is 'catalysed' by the spin-statistics link. This exchange interaction, in its role of favouring aligned spins so as to lower energy, is the basic mechanism of ferromagnetism, a macroscopic magnetic moment arising from the individual moments of the electrons. Similar but opposite effects occur in nuclei where attractive interactions occur between identical nucleons so that an antinode along the 45-degree line now lowers the energy and the singlet coupling of the pair is favoured.

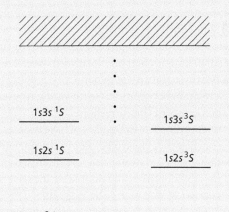

Figure 2.2 The spectrum of the helium atom, He, showing singlet–triplet splittings. The ground singlet, and excited state pairs of singlet and triplet $^{1,3}S$ symmetry are shown.

With increasing excitation, especially double excitation of both electrons in the helium atom (or in any other atom similarly), the 45-
degree line plays even more of a significant role. States in which both
electrons share the excitation equally have wave functions that concen-
trate along that line, called a 'ridge' because of the tendency for the
two-electron system to fall off it, and towards the two axes, that is, for
one or the other electron to become closer to the nucleus. Physically,
this reflects the mutual screening between the electrons of the positive
nuclear Coulomb field. Any particular division of energy between the
two and thereby some ratio of r_1/r_2 is only enhanced as the overall size
of R increases. The slower of the electrons hangs back closer to the nucleus, screening its field for the faster, which only makes it faster still, at
the expense of the inner electron.

In either a time-dependent picture in terms of t or a stationary-
state time-independent picture in terms of R, this 'dynamical screening'
makes the ridge a position of unstable equilibrium. Especially for double
ionization of an atom (say, for instance, by absorbing sufficient energy
either from a photon or some other collision) just above threshold,
when the two electrons have just enough energy to escape to infinity, the escape depends on maintaining roughly equal division of the
small amount of available energy and staying on the ridge for most of
the double-escape process. The same considerations apply to very high
doubly excited states with equal energy sharing between the electrons
that lie just below the double-ionization threshold. In such states, the
electrons have small kinetic energies, albeit with overall negative energy
for the state. What these features of dynamical screening imply is a very
strong radial correlation between the electrons that preserves $r_1 \approx r_2$ in
such states in the vicinity of that threshold.

In terms of the coordinate α in Eq. (2.7), whereas states of single excitation have one or the other electron closer to the nucleus, that is, $\alpha \approx$
$0, \pi/2$, high doubly excited states of the sort discussed in the previous
paragraph and the states of threshold double escape have wave function
concentration near $\alpha \approx \pi/4$ along the ridge line. This coordinate α,
which depends on both electrons and is a 'pair' coordinate rather than
the independent-particle r_1 and r_2, together with the other coordinate
mentioned earlier, θ_{12}, the angle between $\vec{r}_1$ and $\vec{r}_2$, are natural coordinates for describing the states of a two-electron system in which the
electrons are on par. They prove convenient for describing the radial
and angular correlations, respectively, between the electrons so that

transformation to these so-called 'hyperspherical' coordinates (or representation) proves useful. The set (R, α, θ_{12}, Euler angles) describes the two-particle system as a single entity in that six-dimensional space with R the radius of the hypersphere in that space.

Just as in single-particle physics, where an uncertainty link exists between angle and angular momentum as conjugate entities, the more the concentration in one, the broader is the distribution in the other, so also for the pair of electrons. The more concentrated the wave function around the ridge line, $\alpha \approx \pi/4$, the larger is the superposition of harmonics in that angle. Physically, the fact that the angle is a measure of the ratio of the radial distances means that there is a radial correlation between the two electrons. Similarly, the more the concentration is about $\theta_{12} \approx \pi$, that is, with the two electrons lying on opposite sides of the nucleus, the larger is the required superposition of basis states, called Legendre[15] polynomials, in θ_{12}, and the larger is the angular correlation between the electrons. The region around the double-ionization threshold, whether of high doubly excited states or of double escape, when the two electrons are very slow (low kinetic energy), is dominated by $\theta_{12} \approx \pi$, a natural consequence of the repulsion between the electrons driving them to opposite sides. Even when total angular momentum, L, is zero, a large superposition in the individual angular momenta, ℓ, is involved.

The two-electron atom's potential is a sum of the three pair-wise Coulomb interactions, the two attractive ones between electron and nucleus (with charge $+Ze$), and the repulsion between the electrons. Together, we have the potential energy

$$V = -e^2 \left(\frac{Z}{r_1} + \frac{Z}{r_2} - \frac{1}{r_{12}} \right), \tag{2.9}$$

with $r_{12} = |\vec{r}_1 - \vec{r}_2|$ the distance between the electrons. Written in hyperspherical coordinates, it is again a function of three variables, the three dynamical variables. It scales inversely with R and is a function of (α, θ_{12}). This function, $C(\alpha, \theta_{12})$, is sketched in Figure 2.3 for another theme that will occur in Chapter 3 and it shows a saddle point at this

---

[15] Adrien-Marie Legendre, 1752–1833, French. Mathematician with numerous contributions, known for polynomials and transformations named for him, the latter an element of Lagrangian–Hamiltonian mechanics. He also made major contributions to numerical analysis, elliptic functions, and number theory.

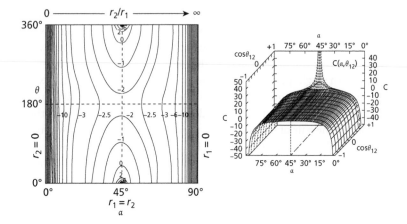

Figure 2.3 Potential surface of a two-electron atom, showing a saddle point in the middle. Contour plot of the potential in Eq. (2.9) at fixed hyperspherical radius R as a function of the angles α and θ_{12} on the left and three-dimensional rendering of half of the potential on the right, the other half symmetrically reflected about the saddle. Adapted from C. D. Lin, *Phys. Rev. A* **10**, 1986 (1974), copyright 1974 by the American Physical Society.

special value of $(\alpha = \pi/4, \theta_{12} = \pi)$. Double escape just above threshold and their associated high doubly excited states immediately below that threshold have their wave functions concentrated in the vicinity of that saddle for most of the range of R as R increases to infinity. This saddle structure becomes a deciding factor in their properties.

Here again is the theme of alternative representations and basis states for describing alternative physics. We may use either the hyperspherical coordinates (R, α, θ_{12}) (plus the three Euler angles) as the radius and five angles of a six-dimensional sphere describing the two-electron system as a whole, or the independent-particle coordinates of two vector directions in three-dimensional space of the two electrons. The two-electron system Hamiltonian, because of the electron–electron interaction, $-e^2/r_{12}$, is separable neither in independent-electron coordinates, $(\vec{r}_1, \vec{r}_2)$ nor in the pair coordinates, nor for that matter, in any coordinate system. (This is what is meant by saying that the three-body problem is not exactly solvable, already even in classical physics, and remains so in quantum physics.) Therefore, quantum numbers corresponding to these coordinates, or basis states in terms of them, have no physical significance.

Meaning is attached only to the quantum numbers corresponding to the operators that commute with the total Hamiltonian, again those of the total spin and orbital angular momentum. All others are approximate quantum numbers and descriptions, any choice between them being a matter of convenience or taste and being different ways of talking about an underlying reality which itself does not depend on those descriptions. States closer to the single-particle description, that is, when the two electrons are on an unequal footing, are best described on the independent-particle basis, requiring a superposition of only a few such basis states. Other states that are more pair-like conform closer to the pair basis and pair quantum numbers, requiring only a few of those basis states. Any state of one basis is a superposition of many states of the other, just as is true of Fourier conjugate pairs such as position–momentum.

The ground state of helium should be seen simply as the lowest ^1S state of the helium Hamiltonian. We might utilize either basis and, in principle, it requires an infinite superposition either way. But the $1s^2$ configuration is the dominant component. Similarly, the first excited state of the same overall quantum numbers ^1S is simply the next higher one with such quantum numbers, $L = S = 0$, and orthogonal to the ground state. Again, while in principle an infinite superposition, this state is dominantly $1s2s$. Independent-particle states of this form as products of single-electron wave functions with individual hydrogenic labels of n and ℓ provide already a good description. For this reason, they are indeed so labelled with these configuration labels (see Figure 2.2). A high doubly excited state, on the other hand, even one with the same overall ^1S character, mixes many independent-particle configurations of $1s^2$, $2s^2$, $3s^2$, ..., $2p^2$, $3p^2$, ..., $3d^2$, etc. The very highest states, just below the double-ionization threshold and the state of threshold double escape, may more nearly be pure pair states, labelled by the quantum numbers for α and θ_{12}, which may be called n_{rc} and n_{ac}, respectively, for strong radial and angular correlations. Because they are associated with angle variables, they are again discrete quantum numbers.

2.2.2 Frame Transformations

The preceding section discussed alternative coordinates, of position or momentum, or of Cartesian versus spherical, as alternative representations of a physical system. There is an even wider context to transformations, again already in classical physics but taking on even wider importance in quantum physics. Consider first the classical context that already arose in Chapter 1. A triad of orthogonal axes constitutes a frame of reference for any observer. Besides translations and rotations relating different frames, Galilean or Newtonian relativity made physicists familiar with 'inertial' frames that are related to each other through a uniform velocity (vector velocity, not just scalar speed) relative to each other. All of mechanics is invariant with respect to such inertial frames, expressing the fact that Newton's equations involve the second derivative or acceleration, not the first derivative or velocity. Thus all inertial frames are on a par, their descriptions equally valid, as far as mechanics is concerned.

The equivalence of inertial frames took on even more significance with the advent of electromagnetism and Maxwell's equations. Indeed, Einsteinian relativity, which extended the equivalence of inertial frames beyond mechanics to all physics, is one of the glorious chapters of our subject and we are familiar with the Special Theory of Relativity's 'Lorentz transformations' (Sec. 7.2) between inertial frames that link the description of space, time, and other physical quantities, including electric and magnetic fields in different frames moving uniformly relative to each other. More general transformations between the coordinates of space and time became the basis of Einstein's General Theory of Relativity, embracing also accelerated frames and simultaneously giving a new view of gravitation in physics.

Lorentz transformations between two frames with relative velocity $\vec{v}$ can themselves be viewed as a kind of rotation but involving a spatial coordinate and time. All rotations have their settings in (two-dimensional) planes and should be viewed as such rather than in terms of an axis of rotation, the examples in Sec. 2.1 being either in the x–y or r_1–r_2 planes. The number of rotations in d dimensions is $d(d-1)/2$, so that in two dimensions there is only one but in three there are three independent rotations (accidentally the same number as the number of axes). In enlarging to four-dimensional space–time, three more 'rotations' are added for a total of six, involving planes containing one of the spatial coordinates along with time. As already noted in Sec. 1.2.4, because time enters somewhat differently in the invariant space–time interval with opposite sign for squared distances, $(c^2t^2 - x^2 - y^2 - z^2)$, these are not real rotations but rather 'Lorentz boosts' connecting (inertial) frames with uniform velocities with respect to each other. In place of the trigonometric sines and cosines, the corresponding hyperbolic functions of sinh and cosh express the equations of Lorentz transformations.

Note the role of c, the speed of light in a vacuum, as the dimensional element to place space and time coordinates together in the invariant interval. In current physics, the numerical value of c itself has been 'defined' as fixed (to nine significant figures), and since time measurement (actually its inverse, frequency) is much more accurate than that of spatial distance, which involves comparing with a standard metre kept at the Paris Bureau of Standards, the latter has been removed as one of the fundamental standards. Time standards based on atomic clocks and the defined value of c replace the historical standard of length.

There is a close correspondence to the discussion in Sec. 1.2.5 of six rotations and $O(4)$ symmetry in the hydrogen atom's spectrum. As noted also there, the difference in sign between time and space coordinates, so that three are not rotations but Lorentz boosts, makes this a symmetry of the Lorentz group $O(3, 1)$, the non-compact counterpart of the compact orthogonal group $O(4)$ that describes four-dimensional rotations. A major implication of non-compactness is of infinite-dimensional representations (an infinity of boosts) unlike the finite-dimensional representations of $O(4)$. Indeed, as observed in Sec. 1.2.5 in the hydrogen spectrum, when one considers not bound but continuum states of positive energy, states of electron plus proton with a kinetic energy at infinite separation that may range from 0 to ∞, they also are infinite in number at any energy (ℓ taking all integer values from zero to infinity) so that they too involve the non-compact extension from $O(4)$ to $O(3, 1)$. Such pairs of groups, one with an index set off by a comma, share many algebraic aspects, such as their dimension (four for both), number of generators (six for both), and the structure of commutators between them, except for a relative minus sign in these 'structure coefficients', and the nature, finite or infinite, of their representations.

Frame transformations take on even wider significance in quantum physics and represent even further extension of the meaning of rotation [18]. Consider, for instance, a many-electron atom or molecule. Each electron has orbital and spin angular momentum but, because of interactions between electrons, they are not individually conserved, only the total for the atom. It is only these total quantities, S of spin and L of orbital angular momentum, that commute with the full Hamiltonian and so lend their labels for designating the states (this is when we neglect spin–orbit interactions, otherwise only the label J of the combined total angular momentum $\vec{J} = \vec{S} + \vec{L}$ enters).

Even in the simplest many-electron example of the helium atom, with just two electrons, there arises immediately the question of alternative ways of combining the individual spins and orbital momenta. Sometimes one pathway, wherein all spins are combined first into an S and all orbital ones into an L, and then the two combined into J, may be relevant for the physics (called LS-coupling), while at other times each electron's s and ℓ, first coupled into its j and then the j values added to give the total J (called jj-coupling), may be more appropriate. These provide again alternative representations with different intermediate angular momenta involved and not all quantities mutually commuting.

And, with alternative representations or descriptions, the question arises immediately of a transformation between them, a so-called $LS \rightarrow jj$ transformation now. Like all transformations between different representations, it is unitary and, given that all the quantities involved are real and finite in number, it is now indeed a rotation, but in an abstract finite-dimensional space of angular momentum coupling rather than in the three-dimensional space around us. With more than two particles, there arise even more possibilities for combining angular momenta, even more representations, and even further transformations. Quantum physics, therefore, expands even further the gamut of frame transformations in physics.

Indeed, it is natural to use such frame transformations even in describing a single physical phenomenon. Thus, consider an event such as photoionization of an atom (or molecule, or their ions) with only a single electron released to infinity, leaving behind the rest of the many-electron system as a positive ion. The photon is absorbed by an electron when it is close to a heavy nucleus in order to conserve energy and momentum. Subsequently, it escapes to infinity. (This is clearly a description sequential in time, although a time-independent description is also possible, a theme of Chapter 7.) It is natural to consider different coupling schemes between it and the other electrons at different stages of this process. When it 'starts' on this escape process and is still close to the other electrons and strongly coupled to them, it is appropriate to adopt the LS-coupling scheme, treating all electrons on a par and combining all their spin and all their orbital momenta first, before adding those totals to get the full angular momentum, J. But when the ejected electron reaches asymptotic distances and is far from the other electrons, it is more naturally described by coupling its orbital and spin momenta together and the resulting j to the total angular momentum of the residual system in jj-coupling. Thereby, the escape process involves in its description of the evolution of the electronic wave function an $LS \rightarrow jj$ frame transformation [18].

2.3 States and Transformations

We deal with states of a physical system and transformations between them. Whether in classical or quantum physics, the two go hand in hand. Indeed, they are so inseparable that it makes little sense to draw sharp distinctions between them, let alone argue for the primacy of one over the other. As in Sec. 1.2.3, even what might appear to be a static constraint can be incorporated instead through a Lagrange multiplier and treated as a dynamical variable. Again, these remarks apply even more in quantum physics. Before considering them, let us also note similar appearances outside of physics.

It is said that the philosopher Whitehead[16] was once asked by students, 'Professor, which are more important, ideas or things?' He replied immediately, 'Why, I would think it is ideas about things'. The same duality between foreground and background appears in art and music. The Richard Strauss[17] opera *Capriccio* is a debate between what is more important, words (or poetry) or music, the composer again seeing the resolution in 'words set to music', appropriately in an operatic form that combines both.

---

[16] Alfred North Whitehead, 1861–1947, English. Philosopher, logician, and mathematician, with major contributions to the foundations of mathematics and the philosophy of science. Emphasized the central role of process in philosophy. Co-author with his pupil Bertrand Russell of *Principia Mathematica* and a major influence on several well-known philosophers.

[17] Richard Strauss, 1864–1949, German. A major music composer of the Romantic and early Modern era, perhaps the greatest composer of the first half of the 20th century. Known for his tone poems, *Four Last Songs*, and several operas.

In quantum physics, the intimate connection between states and operators that act on and transform them is exemplified down to the very notation we use for Dirac kets $|\rangle$ (and their adjoint bras $\langle|$) for states and a ket-bra, $|\rangle\,|\langle|$, for an operator. To emphasize this, let us first look at just the patterns inherent in this notation itself without any reference to what they stand for in physics. Besides bras and kets, quantum physics employs another entity, a multiplication of them, a bra-ket, denoted, naturally, $\langle|\rangle$, this product a number, generally complex. It is these numbers that are ultimately what we observe or measure in our laboratories. This 'bracket' and splitting it into the two entities from which it results as a product is so compelling in itself that, ever since its introduction by Dirac, it has been adopted by physicists as the natural language for quantum physics, along with the coinage of two new words, bra and ket, for the states involved.

With a bra-ket $\langle|\rangle$ representing a number, note from the very structure itself that a ket-bra-ket, $|\rangle\langle|\rangle$, is a complex number times a ket and thus itself another ket (multiplication by a number is the same whether from left or right), representing the state resulting from the action of the ket-bra, an operator (or transformation), on the original ket. Thereby, in just the notational structure of the basic elements of ket, bra, ket-bra, and bra-ket (state, adjoint state, operator, and number, respectively), we can see the common footing and inter-connectedness of states and operators (see Figure 2.4). This is another illustration of the power of a notation that seems natural to the physics, as was observed in Sec. 1.2.4 with four-vectors for relativistic kinematics.

One element involved in adjointness is complex conjugation, so that when a ket is represented by a wave function, the bra involves the complex conjugate of that function. The bra-ket implicitly involves integration over the product of the two functions to give finally a number. When a ket is represented by a column vector, adjointness involves an interchange of rows and columns so that the bra is a row

$$|\rangle$$
$$\langle|$$
$$|\rangle\langle|$$
$$\langle|\rangle$$

Figure 2.4 Elements of quantum mechanics in Dirac bra and ket language. Shown from top to bottom are ket and bra states, operators (ket and bra in that order), and numbers (bra and ket multiplied in that order).

vector. Thus, a bra-ket is the matrix product of a row and a column and, therefore, a number. On the other hand, a ket-bra is a square matrix, an operator that can act through matrix multiplication rules on ket-vectors from the left or bra-vectors from the right to give other kets or bras, respectively.

States, along with their bras and kets, and operators are abstract entities and we do not have apparatus that directly access them. There are no 'wave function metres'. All our apparatus yield numbers, real numbers, and this is what we measure. (This is but a tautology, tied to our usage of the words real and reality!) Since a bracket, sometimes also referred to as a 'matrix element', is in general a complex number, a further crucial step is that experimentally measured quantities are expressed in terms of the squared modulus of the bra-ket. The Born probability interpretation (Sec. 1.2.2) is part of it, wherein the bra is the adjoint of the ket with the unit operator in between.

The ket-bra is thus quite naturally an operator or transformation taking one state to another. Operators and states are intimately tied in the physics of our quantum world. As stated at the beginning of Sec. 2.2, quantum and classical physics differ in the meaning of a physical state. In the latter, the state of a physical system is specified by providing coordinate positions and velocities, themselves observable real numbers, of all the particles involved. This is ruled out by quantum principles, notably the prohibition against specifying simultaneously both the position and the momentum of a particle.

There is, however, a well-defined state of a physical system in quantum physics as well, only that it is specified in one of many ways or representations. Using the coordinate representation, it is a complex wave function, $\psi(\vec{r})$, whereas in the momentum representation it is a different complex function, $\phi(\vec{p})$. In a matrix representation, it is either a row or column vector. More abstractly, we need only designate it as a state $|\ \rangle$, leaving open any particular representation we may choose to work with.

It remains true, however, that a principal aim of physics, or more accurately dynamics, is to give definite predictions of how the state will evolve in time. Just as Newton's equations of motion permit us to follow the evolution in time of the positions and velocities, that is, the state of the system as understood in classical mechanics, so too the fundamental equations of motion in quantum physics. Knowing the potentials involved allows us to follow the time evolution of the state, $|\ \rangle$, or the wave function, ψ or ϕ. In this, quantum physics is as deterministic as is classical physics. We will return in Sec. 8.5 to further aspects of classical and quantum descriptions.

It is just that quantum states are not directly accessible to our measurement. Instead, transformations between different representations, together with all the observables of interest such as position, momentum, energy, etc., are operators $|\rangle\langle|$ that act on states, and what we observe are the numbers resulting from sandwiching such an operator (or products of them, with a final form $|\rangle\langle|$). The final object of interest, and the one connected to our observational or experimental results through perhaps a modulus square of those complex numbers, is always of the form of a bra -. . .-ket.

It is a matter of indifference to physics whether one physicist chooses primarily to work with and use the language of states and another with operators. Sometimes these are called, respectively, the Schrödinger or Heisenberg approaches. States are always involved at the left and right end of the final complex number from which the physics is extracted. In quantum mechanics, since, given a state, others are immediately generated by the action of the myriad operators, there is naturally a whole set of states, or a Hilbert space of states. With the same state labels standing in both, a ket-bra is a projection operator, since acting on any state it produces the ket of that particular state multiplied by a number. Clearly, repetition of a projection operator leads back to the same operator. If the projection operators of all states are summed, that is equivalent to multiplying or transforming by the unit operator, so that the sum is the unit operator. This is referred to as 'closure' and the set of states is said to be 'complete'.

Quantum field theory, which views all physics in terms of interacting fields, deals mostly with operator products but also needs a state called the vacuum state whose bra and ket stand at the two ends of the operator product, the resulting 'vacuum expectation value' containing all the physics. Particles, or more accurately states of many particles, are seen as the result of the field operators acting on the vacuum and exciting these entities that are seen as particles (Sec. 7.3.3).

With this intimately intertwined aspect of fields and particles, the so-called wave-particle nature of all physical systems is also better described as associated with the representation one chooses. The two aspects, particle and wave, are conjugates in terms of the space in which they are localized, coordinate and momentum space, respectively. As conventionally understood in classical physics, a particle is located at a point in space, while a wave has a definite wave vector (or, equivalently, momentum) but is spread out over all space. Instead of 'wave-particle duality', it might have been better to have seen any quantum system as a 'wavicle' characterized only by a wave function, ψ, the underlying reality of a quantum system being in neither coordinate nor momentum space but viewed by either as $\psi(x)$ or $\phi(p)$. Each is sufficient and complete to determine all the physics with neither having any claim to a special footing. It is the 'locators' or the physical apparatus involved that realize one or other representation. These apparatus being classical in nature, on the one hand we may detect charges or other identifying features of an electron, proton, or even neutrino (as a zero mass object) or, on the other hand, electric field amplitudes and phases of electromagnetism. See also Sec. 8.5.

Indeed, as nicely described in [4], given conserved quantities such as charge or non-zero spin angular momentum, electrons or neutrinos have the particle as their classical limit while only for zero mass, uncharged bosons (there are but two examples, the photon and graviton) do we have a wave as their classical limit. It is not surprising, therefore, that physics first made the acquaintance of electrons, protons, neutrinos, charged pions, etc., as particles, and only electromagnetism and gravitation as wave fields (although in the latter, only the monopole or static field has so far been seen directly in experiment, and the detection of gravitational waves is not yet in hand – there is only indirect evidence for them, in the slowing down of the orbits of binary neutron stars).

As further remarks on the use of alternative representations, it is worth noting that there is much to be gained by these different pictures or approaches. While even more widespread in quantum physics, it has also been true in classical physics that different pictures illuminate different aspects and are therefore valuable. The underlying reality that we are trying to grasp always lies beyond our models and understanding, and we can only hope to get closer without actually 'reaching' it. Thus, in what we understand of the state of a physical system, there has been a very big change of ground from classical to quantum descriptions. But in both, different descriptions and approaches to the same problem provide a better approximation to that underlying reality, even when any single representation may be intrinsically capable of a complete description. Therein lies the virtue of seeing the world from different points of view (see also Sec. 8.5).

In this, physics sits in the wider context of intellectual inquiry. Whether in the movie *Rashomon*, or a novel, each description may capture some essence, together getting closer to a full comprehension, closer but not capturing or identical to that reality itself of the complex whole. Reality is . . . what it is. It is said that when Tolstoy[18] was asked to describe what his *War and Peace* is, a novel or history or a historical novel, he said that it is 'not a novel, still less a historical chronicle but what the author wanted and was able to express in the form in which it is expressed'. This could well serve as a paraphrase of how a physicist views physics, that it is the way nature is expressing its underlying reality.

---

[18] Leo Tolstoy, 1828–1910, Russian. Writer of novels that are household names around the world, and whose pacifist and social reform views make him one of the world's great moral thinkers. He had a deep influence on Mahatma Gandhi and Martin Luther King.

3

Localization at Saddles

3.1 Saddles in Terrains and Physics

From the unfortunate Ötzi the Iceman, whose mummified remains were discovered 5,000 years after his death on a high mountain pass in Italy, to hikers of today, saddles in mountainous regions have played an important role in traversing such terrains. We are familiar with the fact that to get from one valley to another, one hikes over an intervening saddle. Unlike at a peak, where the land falls off in all directions, or its opposite of a valley, with climbs in all directions, a saddle point of a surface is one where the ground drops off in some directions while it climbs in other directions. Like peaks and valleys, it is also a stationary point of the surface. That is, the first derivative, the slope, vanishes in every direction, making for local flatness, but the second derivatives do not satisfy the conditions for an extremum, whether maximum or minimum. It is remarkable that saddle points also play crucial roles in physics, both classical and quantum, and that will be the theme explored in this chapter.

3.1.1 Stability at Mechanical or Electromagnetic Saddles

A pendulum, or one-dimensional oscillator, is again a good place to begin, just as we did in Chapter 1. This parabolic potential, $kx^2/2$, with $k > 0$, has a stable minimum (Figure 3.1) with associated small oscillations around it, as discussed for the pendulum. However, a real pendulum's gravitational potential, $mg\ell(1-\cos\theta)$, has a sinusoidal form (Figure 3.2), and reduces to the parabolic form only for small values of θ. That is, only for small displacements from the normal hanging position, vertically down, are there the simple harmonic motions with the time period given in Eq. (1.1).

This potential also has another equilibrium point, at $\theta = \pi/2$, termed the 'inverted pendulum', with the pendulum string (a thin stick serves to represent this situation better) and bob vertically up rather

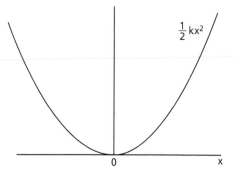

Figure 3.1 A one-dimensional harmonic oscillator's parabolic potential. The spring constant, k, determines the strength of the potential.

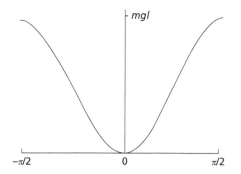

Figure 3.2 The gravitational potential of a pendulum of length ℓ as shown in Figure 1.3. For small oscillations in the vicinity of $\theta \approx 0$, the potential reduces to the parabolic form of a simple harmonic oscillator in Figure 3.1. But at the points $\theta = \pm\pi/2$, which describe positions of the inverted pendulum, there is unstable equilibrium.

than down. While also an equilibrium point in having vanishing slope (zero force), this is now an unstable equilibrium. Even the slightest perturbation will lead the bob to fall away from that position. In terms of the potential, it is an inverted parabola with $k < 0$, and the bob will fall under gravity.

In one dimension, that is all there is to the story. There are only maxima or minima in realistic physical situations even though there are mathematical functions other than quadratic where the second derivative is also zero and only some higher derivative is non-vanishing. But,

Figure 3.3 Saddles in mountain terrains. Paulba Legend <http://en.wikipedia. org/wiki/File:Washington-clay_saddle.JPG>.

as soon as one goes into more than one dimension, even just a two-dimensional surface, saddles can arise besides peaks and valleys, just as they do in the topography of the Earth's (two-dimensional) surface (Figure 3.3). New forms of dynamics now become possible.

A striking example is provided by Figure 3.4, which shows the simplest such situation of a surface with just one saddle. There is one direction of stable motion and one of unstable motion away from the saddle. This is as simple as one can get, with just one motion of each kind. A marble placed at the exact centre of the saddle upon experiencing the slightest disturbance along the unstable direction will fall off the surface. In that direction, this situation is exactly analogous to the inverted pendulum. But now imagine introducing a time element by placing the saddle surface on a turntable, as indicated in Figure 3.4, that is spinning around the vertical axis. Upon spinning with sufficient angular speed, the marble can be stabilized. The dynamical problem exhibits a stability that is not there in the static potential. Indeed, there is also a one-dimensional counterpart, an inverted pendulum's instability compensated when the point of suspension is jiggled up and down or by some alternative feedback mechanism, as one knows by balancing a long umbrella with its tip on one's finger (your eyes have to focus on the umbrella handle and the eye–brain feedback work to keep the umbrella vertical).

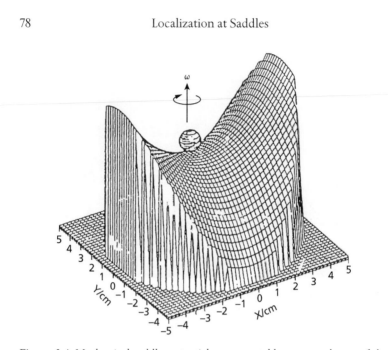

Figure 3.4 Mechanical saddle potential on a turntable as an analogue of the Paul trap for electrical charged particles. A quadrupolar electric field, the analogue of the saddle shown, will not itself hold a charged particle, the analogue of the marble. But, in combination with time-varying radio frequency electric fields, the analogue of the rotation of the turntable, the particle can be stabilized. From W. Paul's Nobel Prize Lecture, *Rev. Mod. Phys.* **62**, 531 (1990), copyright 1990 by the American Physical Society.

The reason for this stability is the Coriolis[1] force, which acts on moving objects perpendicular to their velocity and to the angular velocity. (There are many manifestations of this force in atmospheric and oceanic swirls and storms because of the Earth's rotation.) Thus, as the marble falls in the unstable direction in Figure 3.4, this Coriolis force points parallel to the stable direction and, in turn, leads to a restoring force in the unstable direction driving back to the equilibrium point. The net result is a rotating force that keeps the marble around the equilibrium point. A time-dependent field analogous to the one here

---

[1] Gaspard-Gustave Coriolis, 1792–1843, French. Mathematician and mechanical engineer who was interested in applied aspects of work and energy in machines and especially in water wheels, and thus rotational energy.

from spinning also can stabilize a one-dimensional inverted pendulum, as noted. In that case, jiggling the point of suspension up and down with a high enough frequency can keep the pendulum vertical.

Celestial mechanics has long recognized so-called 'Lagrange points', which are precisely such quasi-stable positions in the Sun–Earth (or Earth–Moon) system. Points roughly 1,000,000 km on either side on a line perpendicular to the one joining the Sun to the Earth are saddles in the gravitational potential of the two bodies. While these would be unstable points were all the bodies static, because of the rotation in the system, the Coriolis forces give quasi-stability and, indeed, man-made satellites have been parked there. Nature itself has done the same with the so-called 'Trojan asteroids' at similar Lagrange points of the Sun–Jupiter system. Similar examples are known for other planets.

There is also an electromagnetic analogue of the saddle, the so-called 'Paul[2] trap' for trapping positive ions. Indeed, Figure 3.4 and the mechanical analogue were presented in the Nobel Prize lecture of Wolfgang Paul as a mechanical model of his invention for trapping charged particles. It is well known that a charge cannot be stably held with purely electrostatic fields (except trivially on top of an opposite charge). Such fields have to satisfy Laplace's equation, which stipulates that at a point where there is no charge, the sum of the three second derivatives of the electric potential with respect to (x, y, z) has to vanish. Therefore, at least one has to be of negative sign, that is, an inverted parabolic potential, which corresponds to unstable motion in that direction for any charge placed there. Paul's discovery was to place in addition a time-dependent radio frequency field besides the quadrupole fields that gave trapping in two directions to get overall dynamical trapping in all three dimensions. Such traps have been enormously influential. Another solution, and thereby another class of traps, is to combine electric and magnetic fields; these are Dehmelt traps. Paul and Dehmelt[3] shared the Nobel Prize in Physics.

---

[2] Wolfgang Paul, 1913–93, German. Physicist who invented a way of trapping positively charged ions. He opposed the deployment or use of tactical nuclear weapons by the West German Army.

[3] Hans Georg Dehmelt, 1922, German and American. Developed methods for trapping charged particles and made precision measurements of magnetic moments and g-factors of electrons and positrons.

3.2 Saddles in Quantum Systems

As in classical mechanics or electromagnetism, since quantum physics also deals with motion in potentials, not surprisingly saddle points that arise when there are two or more degrees of freedom are also important for quantum systems. Indeed, in all these problems of classical or quantum systems, clearly saddles proliferate with increasing dimensions. It is less likely that all directions will have the same sign of k, whether positive or negative, but more often a mix, so that some directions away from equilibrium will be stable and others unstable. Therefore, the understanding of physics in saddles rather than at a global maximum (peak) or minimum (valley) takes on added importance with more degrees of freedom. This understanding is best gained by considering the simplest example, in which there is only a single saddle in a two-dimensional problem, and we consider two such systems (Sec. 3.2.2 and Sec. 3.2.3). But first we consider localization more generally in quantum systems.

3.2.1 Localization in Quantum Systems

In classical physics, the concept of localization of a particle is simple, when it occupies a particular position in space. Thus, in one dimension, say in a potential well, as shown in Figure 3.1, the minimum energy of a particle of mass m is when it sits at the bottom of the potential, at $x = 0$, where both kinetic and potential energy, and thereby total energy, equal zero. States of higher energy, when the particle can rattle around in the well, will, in realistic situations, with dissipative forces such as friction, gradually settle down to the minimum energy, with the particle coming to rest at the bottom.

In quantum physics, such a configuration is forbidden, because simultaneous position and momentum at definite values, as in the particular case when both are zero, is not allowed by the uncertainty principle. Thus, a quantum particle in the parabolic potential well will not reach zero energy and $x = 0$ as its lowest state but rather have a 'zero-point energy', $\hbar\omega/2$, with frequency $\omega = \sqrt{k/m}$. Its wave function, shown in Figure 3.5, is a Gaussian function that exists over the entire space so that the probability of its location, given by the square of the wave function (Secs 1.2.2 and 2.2), is non-zero everywhere but is peaked around $x = 0$. Note, in particular, that there is a non-zero probability of finding the particle in the so-called 'classically forbidden zone' beyond the potential

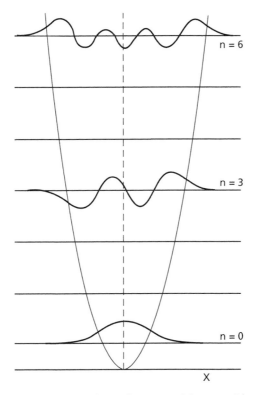

Figure 3.5 Energy positions and wave functions of the ground ($n = 0$) and two excited ($n = 3, 6$) states of the one-dimensional quantum harmonic oscillator's potential well of Figure 3.1. Note the Gaussian tails of the wave functions at $\pm\infty$, number of nodes n, and parity symmetry/antisymmetry for n even/odd. With increasing n, the peak value of the wave function occurs closer to the classical turning points where the energy, E, equals the potential energy.

walls at that energy, that is, with the potential energy higher than the total, seeming to imply a negative kinetic energy. Of course, such classical thinking does not apply, it being incompatible to talk of energies in regions of space. What remains true is that the spatially averaged kinetic energy is never negative, even in quantum systems.

Although not localized at a single point, nevertheless the wave function may be so peaked as to have appreciable probability confined to a small region around $x = 0$, depending on the value of k. The same holds true for other forms of potential besides the parabolic one for a harmonic oscillator shown in Figures 3.1 and 3.5. The deeper the potential well, the more localized the ground state. Also shown in Figure 3.5 is the wave function of a higher energy state. Again, in a quantum oscillator, unlike in a classical one of the same frequency, not all possible energies are allowed, only the discrete values $(n + 1/2)\hbar\omega$, with $n = 0, 1, \ldots$. The wave function for such a higher energy state now has many oscillations, given by the quantum number n, indicating the number of zeroes, called 'nodes', which are positional values with zero probability of finding the particle when it is in such an energy state.

At large quantum numbers, the probability peaks further away from $x = 0$ and, indeed, close to what is termed the classical turning point, where the total energy equals the instantaneous potential energy. This is plausible because classically it is at these points that the velocity vanishes and the particle turns around in its motion (the end points of the swing of a pendulum) and one expects the probability of finding a particle to be inversely related to its speed. This is an illustration of what is termed the Correspondence Principle, which states that the classical limit of quantum physics is often reached at large quantum numbers. This is obvious in the case of angular momentum, which is always quantized as $\ell\hbar$, and any classical value, however small, necessarily involves very large values of ℓ given the smallness of $\hbar$. But for other physical properties as well, generally at large quantum numbers, the system is quasi-classical.

Strict localization, while different from the above for wave function concentration, is also possible in quantum physics (at least in non-relativistic quantum mechanics, relativistic field theories being a different story: Sec. 7.3.3). A particle strictly localized at $x = 0$ would be described by what is called a Dirac delta-function, $\delta(x)$, a function defined to have vanishing value at all x other than $x = 0$. With position and energy being incompatible quantum operators, in a reverse of what was said earlier of energy states existing at all values of x, now such a localized particle cannot have any definite energy but must be a superposition of all the energy states of the system. This is true for any physical system, an oscillator or pendulum included. A quantum pendulum can be localized at $x = 0$ or, for that matter, at any specific value of x but at the expense of superposing a large number of states, including of large energies.

For three-dimensional counterparts such as the hydrogen or any other atom, almost classical states, such as the electron in a particular

Bohr orbit, can be made today in our laboratories by exciting with broad-bandwidth lasers many so-called Rydberg states, states with large quantum numbers.

Another form of localization, termed 'Anderson[4] localization', may be pictured as a one-dimensional example. A random potential, $V(x)$, with many minima of random location and depth, can lead to trapping of a particle at the deepest of these wells. Such phenomena have been seen in a variety of situations. But we turn next to a different type of localization in quantum systems, with nothing random about the potentials involved, but rather simple, well-defined ones, such as the Coulomb potential. Nonetheless, a dynamical localization takes place around saddle points of the potential surface, two or more dimensions of course being necessary to produce such saddles and localization. It is this form of dynamical localization that relates to the classical and electromagnetic counterparts discussed in the previous section and shares with them the theme of saddles.

3.2.2 Two-Electron Atoms, States of High Excitation

A quantum system with a saddle in its potential surface has already been introduced in Chapter 2. As illustrated in Figure 2.3, the total potential of a two-electron atom, which depends on an overall scale variable $1/R$ and two angles, as in Eq. (2.7) and Eq. (2.8), can be represented as a two-dimensional potential surface in those angles. This surface has deep valleys at $\alpha = 0, \pi/2$ that correspond to either one of the electron–nucleus distances vanishing and thus the attractive Coulomb potential reaching $-\infty$. On the other hand, the surface has steep peaks at $(\alpha = \pi/4, \theta_{12} = 0, 2\pi)$, when the two electrons are equidistant from the nucleus and lie on top of each other, a configuration of infinite repulsion.

Besides these minima and maxima, the other singular point of this surface is the saddle at $(\alpha = \pi/4, \theta_{12} = \pi)$. This corresponds again to equal distances, $r_1 = r_2$, but now with the electrons on opposite sides of the nucleus. In physical terms, the saddle can be seen as follows. For departures of the electrons from being exactly on opposite sides, that is, for the angle θ_{12} away from π, the repulsion between the electrons

---

[4] Philip Warren Anderson, 1923, American. Noted contemporary condensed-matter physicist, known especially for his work on superconductivity and magnetism, and for his writings on the philosophy of science and emergent phenomena.

drives them back to that value. Hence, this is a stable direction, with the potential surface rising away from that point of $\theta_{12} = \pi$. But, for departures from exact equality of the two distances, that is, for α away from $\pi/4$, the opposite happens, with the potential surface dropping away, making it an unstable direction.

This is a reflection of 'dynamical screening', one electron screening part of the (positive) nuclear charge for the other. As a result, any departure from exact equality means that the electron that is closer to the nucleus screens it more for the other, which, as a result, moves further away from the nucleus. Departures of r_1/r_2 from unity thus get accentuated by the very nature of the Coulomb interactions [18]. Thereby, the two-electron atom or quantum three-body system of a positively charged nucleus and two electrons provides the simplest quantum example of a potential with just one saddle point, one motion stable and the other unstable.

Most discussions of a two-electron atom such as helium that deal with the ground or low-lying states of excitation of one electron concern states lying in the valleys of the potential in Figure 2.3. With the electron–electron interaction energy generally smaller (approximately 15%) than the energy of attraction to the nucleus, these states are amenable to a variety of 'perturbative' techniques that have been extensively developed since the early days of quantum mechanics or to 'variational calculations', which work well for low-lying states in the spectrum. However, an interesting class of excitations called 'doubly excited' states, wherein both electrons are excited, and which lie much higher in energy, require a different understanding.

In the helium atom, all singly excited states lie below 24.6 eV from the ground state $1s^2\,^1S$, the lowest of them at 20.2 eV. Beyond the 24.6 eV 'ionization' energy, one electron is ejected into the continuum and the atom is ionized, that is, left behind as a positively charged ion. The doubly excited states, on the other hand, lie between 60 eV above the ground state and 79 eV, which is the energy when both electrons are ejected, marking the beginning of the double-ionization continuum (Figure 3.6). With both electrons escaping to infinity, the doubly charged bare helium nucleus, called an alpha particle, is left behind.

Since even the lowest doubly excited states, with both electrons lifted from the ground principal quantum number $n = 1$ to the next, $n = 2$, lie above the single-ionization threshold of 24.6 eV, all these doubly excited states are unstable with respect to one electron dropping back to a

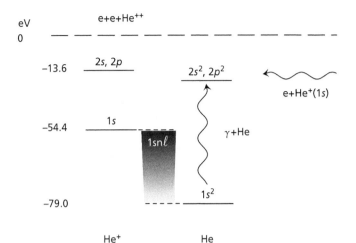

Figure 3.6 Doubly excited states of He. The figure shows the energy levels of the He$^+$ ion at values $-54.4/n^2$ eV. These are the Bohr values for a single electron bound to a nucleus of charge $+2e$, and they converge to the double-ionization threshold (chosen as the zero energy value) with two electrons separated to infinity from the bare nucleus of He^{++}. Below the He$^+$ ($n = 1$) lie the singly excited states of He, as in Figure 2.2, whereas below He$^+$ ($n > 1$) are the doubly excited states, both electrons excited away from the ground $n = 1$ configuration. These doubly excited states are degenerate in energy with continuum states built on lower n states of the He$^+$ ion, decaying into them by the process of autoionization. They may be excited, as shown, either by absorption of one or more photons from lower states of He, or by electron impact on the He$^+$ ion, or through other mechanisms.

lower value of n, as the energy released is sufficient to allow the other to escape to infinity. This process wherein one electron is ejected is called autoionization and is inherent to the Coulomb potentials inside the atom, and can happen on its own even in the absence of any coupling to an electromagnetic radiation field (as needed for the decay of singly excited states). Some of the low doubly excited states have autoionization lifetimes of 10^{-12} s, much shorter than the typical radiative decay lifetime of 10^{-8} s of a singly excited state to a lower energy state. (The four orders of magnitude represents the square of the fine-structure constant, an index of the strength of the electromagnetic interaction, intensities and lifetimes scaling quadratically with the strength.)

Doubly excited states and those in the double continuum display new physics not seen in the singly excited/ionized domain. One such new feature is that, even though energetically allowed to decay, some states have long lifetimes before autoionization, indeed so much so that they descend to lower states by radiative decay, emitting a photon, rather than by ejecting an electron. Interestingly, it is these states that are associated with the saddle in Figure 2.3. Their wave functions, while of course spread over all space, are dominantly confined to the region around the saddle. This saddle lies high above the low-lying singly excited states in the valleys whose wave functions are of course mostly confined there. The overlap between the two classes of states is small and thereby also any transition between them that is governed by the matrix element of the interaction operator $1/r_{12}$ between the bra of one and the ket of the other.

Such a dynamical localization of quantum states into saddles is an interesting theme of many-particle systems. While lying high in energy above other states (sometimes multiply infinite in number), that localization permits a quasi-stability and long lifetimes. Since the saddle corresponds to the two electrons being on a par in their radial distance from the nucleus, a particular subset of doubly excited states with approximately equal excitation of both electrons can be distinguished. Such states lying just below 79 eV above the ground state of helium are also closely related in their physics to the states of the double-ionization continuum on the other side of that threshold.

Indeed, when the helium atom absorbs energy just above that threshold, while it is energetically possible for both electrons to be ejected, the dynamical screening described at the start of this section makes it imperative that, for most of the escape process, the two share the small excess energy available equally in their kinetic energies. Otherwise, should one get more, it will only get faster relative to the other that is hanging back and, in turn, will screen further the nuclear attraction for the outer electron. Finally, only one electron will escape, the other falling back into a bound, singly excited state. That is, the configuration will end up in one of the valleys rather than staying at the saddle out to large R and all the way to infinity which is necessary for both electrons to be ejected. Thereby, double ionization just above threshold requires staying near the saddle in Figure 2.3.

The unstable direction, α, thus plays a crucial role in high doubly excited states and threshold double ionization [18]. This α coordinate being a measure of the ratio of the radial distances, instability in it is an aspect of radial correlation between the electrons. As in any quantum pair of conjugate variables, confinement in one, here in the angle α to a small region around $\pi/4$, means a superposition of a large number of the conjugate harmonics (Sec. 2.2). Likewise, the stable direction θ_{12}, confined to around π, translates into a large superposition of the corresponding spherical harmonics in the conjugate variable ℓ in describing such localized states of double excitation or ionization.

The total angular momentum commutes with the total Hamiltonian of the three-body system, and is, therefore, conserved. Its value, thereby being a 'good' quantum number with a definite value, may be small, even zero. These remarks apply even for such states of ^1S (total orbital and spin angular momentum zero) around 79 eV. They also do, of course, to states with higher orbital and spin angular momentum of the pair.) It is the ℓ of the individual electrons that can and must embrace large values even when the sum remains small (of course, for S states with zero total orbital angular momentum, the two ℓ have to be equal). This represents a high angular correlation alongside the radial one. Today, many experimental measurements are available of these strong correlations in doubly excited states and threshold double ionization, even in the simplest case of two-electron systems. Of course, multiply excited states also display such correlations and, as noted, saddles proliferate with increasing numbers of particles.

An earlier theme from Chapter 2, of alternative representations, also applies to this picture. Independent coordinates of the two electrons (see Sec. 1.2.5), $(\vec{r}_1, \vec{r}_2)$, or their associated quantum labels (n_1, n_2) and (ℓ_1, ℓ_2), serve well to represent low-lying states in the valleys of the potential but become a poor choice, requiring large and unwieldy superpositions of them, to describe the highly correlated saddle states around the double-ionization threshold. They are more appropriately viewed in an alternative representation of 'pair' coordinates (R, α, θ_{12}) and corresponding pair quantum numbers. Of course, the reverse is also true, that it would take a large superposition of pair states to describe a singly excited state in which the electrons are far apart with little correlation between them. The two are alternative representations and, as with all such, both are complete sets and therefore each is capable of describing the physical system of the two-electron atom. The question as described in Sec. 2.2, as always in quantum physics, is the suitability of one or the other alternative representation, depending on the class of states or phenomena being studied.

3.2.3 Transition States in Chemical Transformations

Chemistry involves the motion of electrons and nuclei in an assemblage of atoms. The previous sub-section has considered saddles in the motion of two or more electrons. For chemical transformations, this may be supplemented by considering the motion of the heavier nuclei. Again, as a paradigmatic example, three bodies/atoms in a rearrangement AB + C $\rightarrow$ A + BC can be a stand-in for the general chemical transformation so crucial throughout chemistry and much of biology.

It has become increasingly clear [20] that a transformation such as this is often mediated by a so-called 'transition state' of the combined entity (ABC) that acts as an intermediate, AB + C $\rightarrow$ ABC $\rightarrow$ A + BC. Further, one can view the initial and final configurations as states in valleys of a full potential surface of the system (with many more degrees of freedom than with the two-electrons considered previously) with the transition state residing in the saddle separating them. The three-body intermediate in the saddle may have wildly different lifetimes, some very transient, others metastable enough to show up as resonances at definite energies, but the idea itself of an intermediate playing a catalytic role in the transformation is well established. The calculation of the full potential surface and of transition states, at least for small molecules, has now reached a fair level of sophistication in computational quantum chemistry [20].

3.2.4 Coupling to Another Dimension for Stability

Finally, consider the connection between the quasi-stability of the quantum systems in the previous sections, and the stabilization at saddles of mechanical and electromagnetic classical systems of the earlier Sec. 3.1.1. Those were phenomena with explicit time dependence. The static potential alone with a saddle does not give stability. It is the addition of a time-dependent element, whether a jiggling of the point of suspension or a rotation or a radio-frequency field, that provided the crucial element for understanding stability in terms of the resulting Coriolis-like forces. But the two-electron problem's doubly excited or threshold continuum states are stationary states of a time-independent Hamiltonian. Although in some of the words used of dynamical screening, a time sequence of a series of successive snapshots of the two escaping electrons may have crept in, the analysis itself considers only the time-independent Schrödinger equation (this relates to another theme about time that will be considered in Chapter 7).

Where then is the connection between these problems, some time dependent and others time independent? The answer lies in recognizing that the crucial element is a coupling to another variable, whether it be time, t, or size, R, in the two cases. Time as such is not of the essence; rather, it is the presence of another variable or coordinate that is the important element [19]. Indeed, in the detailed analysis of wave function localization around the saddle, besides the (α, θ_{12}) involved in that potential, a crucial term comes from a piece of the kinetic energy operator in R that is linear in the first derivative in R. This is analogous to the Coriolis force in the mechanical and electromagnetic problems, also stemming from kinetic energy, that involves the angular velocity, similarly a first derivative but there in t.

4

Coins, Classical and Quantum

4.1 Coins in Classical Language and Physics

The coin has many metaphorical uses in our ordinary languages: coin of the realm, false coin, bad coin, etc. Its main feature of two valuedness, heads or tails, is something we become familiar with from early childhood (Figure 4.1). Together with the (electric) switch (Figure 4.2), another term ubiquitous in the language, it then becomes a stand-in for any two-valued property: up/down, on/off, in/out, yes/no, male/female, etc. Thus, it can be said that 'virtue on one side may appear as vice on the other'! The tossing of a coin has become the ultimate in unbiased choice between two alternatives, and repeated tosses seen as generating randomness.

In our age of computers, the mathematical rendering of a coin in binary terms as 0/1, along with the logical true/false, the two values of a 'bit', has become the basis of all the computers and ancillary devices

Figure 4.1 An early coin, heads/tails representing classical two valuedness. This is a coin from the 1st century BC, from a Celtic people called the Veneti, who lived on the Brittany peninsula. <http://en.wikipedia.org/wiki/File:Veneti_coin_5th_1st_century_BCE.jpg>.

Figure 4.2 An electrical switch, a stand-in for two-valuedness: on/off ≡ 0/1. Jason Zack <http://en.wikipedia.org/wiki/File:On-Off_Switch.jpg>.

that run our world. Electrical or electronic switches are the millions and billions of bits in any of these devices or instruments.

The coin runs as a metaphor throughout physics as well. An apt example occurs in a feature of our physical world, that it seems to have electric charges of either sign that can occur independently of each other; but this is not so with magnetic poles, all magnets being a N/S pole tied together, a 'dipole'. 'Magnetic monopoles', although easily accommodated in our physics (as, for instance, into Maxwell's equations in Figure 1.10), seem not to occur in nature, and we have placed stringent experimental and observational limits on their existence. A ready explanation is to see magnetism as always derived from charges in motion, electric currents. The basic current loop acts like a magnet in generating (or reacting to) a magnetic field but, as with any loop or coin, the two faces, north and south poles in this instance, are inextricably tied together. Just as every coin has two sides, so does every magnet have two poles.

4.2 The Quantum Coin

Quantum physics also deals with two-valued entities or two-level systems, the physical system being realized in terms of a basis of just two
states. An excited and a ground state, whether in an atom, nucleus
or any physical system, affords an example. The intrinsic spin angular
momentum of elementary particles such as electrons and nucleons is
yet another, and important, example. It was the advent of quantum
physics, first with the realization that all angular momenta occur as
multiples of the Planck constant, $\hbar$, and next that these multiples are in
either half-odd integer or integer values, that pointed to the importance
of the lowest such value of 1/2 as the basic or 'fundamental' representation of a non-zero angular momentum in our Universe. That there are
many elementary particles, such as electrons, protons, and neutrons,
with a spin of 1/2 is a feature of our Universe. Each such spin can have
two states, with spin projection $\pm 1/2$, commonly called up/down, on
any axis (Figure 4.3).

The spin metaphor itself extends to the observation that for nuclear
interactions the neutron and proton behave as if they were two sides

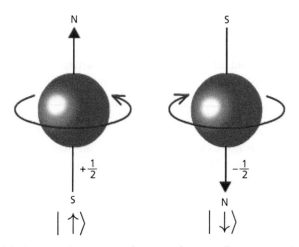

Figure 4.3 A quantum spin shown schematically as a clockwise/
counterclockwise spinning sphere, and oppositely oriented magnetic
moments. Also shown is Dirac ket notation of an up/down arrow for the
two states. <http://chemistry.tutorcircle.com/inorganic-chemistry/quantum-
numbers.html>.

of a coin. Borrowing a prefix from the word 'isotope', used for nuclei with the same number of protons but different numbers of neutrons, this leads to invoking a concept called 'iso-spin'. One object, the 'nucleon', with iso-spin 1/2, has two states, the proton and neutron as $\pm 1/2$, respectively. As with all angular momenta in quantum physics, any angular momentum j (in units of $\hbar$) contains $(2j + 1)$ states, with its projection on any axis itself quantized to take values in unit steps from $-j$ to j (see Sec. 1.2.5).

It is important to distinguish the spin of angular momentum from iso-spin; both reside in abstract spaces but they have different abstract spaces. The helium nucleus comes in two isotopes, one ^{3}He, with two protons and a single neutron, and the other, the more common isotope, ^{4}He, called the alpha particle, with two protons and two neutrons. Their spin angular momentum is 1/2 and 0, respectively, but the two isotopes together may be seen as an iso-spin 1/2 doublet, as two faces of such an isotopic coin. Other particle sets, such as three kinds of pions that can be regarded as three states of iso-spin 1 (while their angular momentum spin is zero), and further connections that this picture provided between processes and decays of elementary particles, made this concept of iso-spin a very fruitful one.

Such two-level systems with two basis states can be said to be quantum coins. While measured always as spin either up or down, or state either excited or de-excited, the quantum coin is intrinsically different from and richer than a classical coin. The basic reason is the linearity of quantum physics and thereby the existence of a superposition principle with, further, the feature that complex elements characterize quantum physics. Any linear superposition of the base states, $|0\rangle$ and $1\rangle$ in Dirac notation, with arbitrary complex coefficients, that is, the state,

$$c_1 |0\rangle + c_2 |1\rangle, \tag{4.1}$$

with complex numbers c_i normalized to unity, $|c_1|^2 + |c_2|^2 = 1$, is also a legitimate state of the system. There is, therefore, a three-parameter (two complex numbers with a real constraint) family of states, far larger than the two states of a classical coin. This is a primary reason why a quantum bit, 'qubit' for short, has more potential than does a classical bit, either for speeding up calculations or for more memory storage. In this, we have the basis of today's fields of quantum computation, cryptography, and teleportation, collectively called the sub-field of quantum information in physics [21].

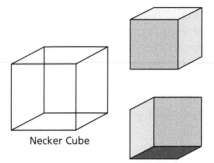

Necker Cube

Figure 4.4 A Necker cube, often presented as an optical illusion but representative of neural image processing, the brain seeing it in two forms, 'in/out' shown on the right. <http://www.optical-illusion-pictures.com/ambig.html>.

An analogy to thinking of superposition is the optical 'illusion' (neural phenomenon, really) called the Necker[1] cube (Figure 4.4). The cube always appears in one of two alternative appearances, as shown in Figure 4.4. Once observed as either, persistent staring freezes in that position but when one takes the eye off and returns, it is possible to see it suddenly 'flipped out' into the second version. In a sense, in between, the cube is in neither definite position, and the act of observation freezes it into one or the other, just those two possibilities and none other. As with any analogy, this should not be pushed too far but serves to illustrate some aspects of a quantum coin and its two observed states, but quantum physics has the added aspect of a very large superposition in the sense of Eq. (4.1).

4.2.1 The Quantum Coin as the Square Root of a Switch

A very fine pedagogical illustration of both the power of the superposition principle and its constraints, and which illuminates fundamental aspects of quantum systems as well as their difference from classical ones, is the random toss of a quantum coin. Consider first a classical coin and a switch, the latter also described as the logical NOT gate. That is, a switch changes between the 0 and the 1 value. In the language of

[1] Louis Albert Necker, 1786–1861, Swiss. Crystallographer, geographer and mountaineer.

game theory and a payoff matrix, this simplest of operations may be described as at the top of Figure 4.5. A classical coin, on the other hand, has the next depiction in Figure 4.5, the output of a toss either 0 or 1, with probability 1/2 regardless of the input.

Consider instead a quantum coin. It too, after a toss, reveals either of the two states with equal probability, but in terms of amplitudes. The payoff differs from that in Figure 4.5 in carrying square roots, and a crucial minus sign or phase for one entry as shown in Figure 4.6. Also, in accordance with the Dirac notation (Sec. 2.3), we have used kets to represent this quantum system. In quantum physics, all results are realized as probabilities based on squared moduli of complex amplitudes. For a single quantum coin toss, the result which depends on such a (modulus) square of the amplitude is just the same as a classical coin's, the square roots and minus sign disappearing under squaring.

Imagine now a sequence of two of these elementary operations. With a switch, repeating it simply reduces to the unit operation, that is, no operation at all, as is obvious from two successive operations in Figure 4.5. With a classical coin, two tosses still lead to the final random

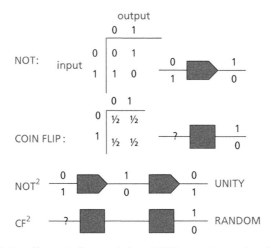

Figure 4.5 Payoff matrix for a switch or NOT operation and a classical coin flip (CF) or randomizing operation, along with the result of two successive applications. Note that a squared switch is equivalent to multiplication by unity (no change of input) but two successive coin flips remain a random outcome. From B. Hayes, *Am. Sci.* **83**, 304 (1995).

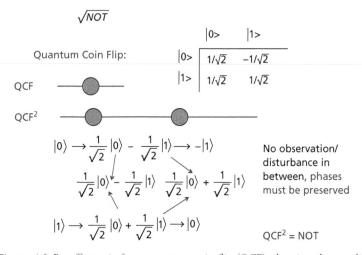

Figure 4.6 Payoff matrix for a quantum coin flip (QCF), showing the amplitude for output ket for a given input ket. Observed quantum coins will have probabilities of outcomes that are the squares of these amplitudes, coinciding with results for a classical coin (Figure 4.5). The operation of a squared quantum coin flip is, however, different. With no observation in between the flips, tracing through either input shows that the squared operation is equivalent to the NOT operation, making a quantum coin flip the square root of NOT. From B. Hayes, *Am. Sci.* **83**, 304 (1995).

result in Figure 4.5, again regardless of the input. But the quantum result is interestingly different. Two successive tosses with no observation in between, this being a crucial element of quantum physics — that observations can change the state — means tracing through two successive applications of the payoff matrix in Figure 4.6. If the input is $|0\rangle$, at the intermediate time between the tosses, the state is in the linear superposition

$$(|0\rangle + |1\rangle)/\sqrt{2}, \tag{4.2}$$

a special case of Eq. (4.1). Any observation at this stage will yield the two base states with equal probability, the state in Eq. (4.2) 'collapsing' into either $|0\rangle$ or $|1\rangle$.

But, without any such intervention, repeating as a successive quantum coin toss, we can trace this state again according to the matrix in Figure 4.6, each of its components itself now becoming a superposition of both. With a little algebra, the net result of the successive tosses is, as shown, the state $-|1\rangle$. An observation will now of course give uniquely the state $|1\rangle$, the minus sign irrelevant upon squaring, so that input $|0\rangle$ results uniquely in output $|1\rangle$. It is a simple matter to show a similar result if one were to start with $|1\rangle$, that uniquely one ends with $|0\rangle$. In summary, the two successive tosses, or the squared operation of a quantum coin flip, is just the same as a switch!

Thus, while the square of a classical coin is the same object, a classical coin, the square of a quantum coin is the operation of a switch (again, at the level of amplitudes, a crucial minus sign must be noted if other operations follow) (Figure 4.6). Inverting this statement, the square root of a switch is a quantum coin flip! Within the realm of classical operations, it is non-trivial and difficult to construct such a square root operation but clearly not so in quantum physics. A single quantum coin provides it. Note the critical role of superposition for the argument in the previous paragraph, the linearity of quantum mechanics being essential.

Caveats are also worth noting. The same linearity that allows an operation that is difficult for classical physics can, in other places, give an advantage in the reverse direction. A familiar example in quantum information is the so-called no-cloning theorem [21] that the very linearity of quantum physics prohibits the construction of a general-purpose apparatus that will reproduce an arbitrary (that is, a superposition) quantum state. By contrast, photocopiers are ubiquitous in the classical world!

Also important in this analysis is that no disturbance takes place between the two tosses. The exact cancellation of one to leave behind only the other base state would be disturbed were the two parts in Eq. (4.2) to encounter different multiplying factors, even pure phases of unit modulus square. In that case, an input base state will be realized in the output as a combination of both base states, with some multiplicative coefficients. Either of the two states is then observed as output, albeit with different probabilities, and we would not have a switch or NOT gate.

The importance of phase is no surprise, given that quantum physics is built on complex elements. It was important in the discussion about the squared quantum coin flip in Figure 4.6 that phases, which are delicate, are not disturbed between the two flips, or, for that matter, at that second flip, to ensure the exact cancellations that lead to an unambiguous pure state of $|0\rangle$ or $|1\rangle$ at the end and not a superposition. This analysis also is instructive on the nature of a two-level system and of measurement in quantum physics. The very premise of a system with only two base states is that every observation of it has only one of those two outcomes. One never measures a superposition as in Eq. (4.1), and all the phase information in the complex coefficients, c_i, is lost.

Whatever the system, whether a coin, or an excited–ground-state pair, or a decaying nucleus, the initial preparation and final detection are of only two possibilities. Either one see the initial nucleus or its decayed products. There is no question of observing it in some limbo state in between. Through quantum interactions in between, such as that at the quantum coin flip, one can create superpositions and transform between them but an observation at the end, with an apparatus, always means a scrambling and loss of information of phases, realizing one or the other base states. The $|c_i|^2$ appear as the probability of which one appears. (In any future quantum computer built of many qubits, it will be true that the initial preparation sets each qubit of a register into one of the base states and the final observation is again when all are in a definite set of base states, although in between and during the process of computation they will explore the large parameter space of superpositions.)

There is a famous formulation called the 'Schrödinger cat', familiar even outside the realm of physics, that was advanced by one of the founders of quantum mechanics seemingly to show the difficulties of the probability interpretation. Schrödinger posed a thought experiment where the radioactive decay of a nucleus is amplified to affect a cat that is enclosed in a box together with the radioactive material. At heart, the physics question is the one at the level of the decaying nucleus as a two-level system but made more dramatic by regarding the two states of the cat as dead or alive, depending on whether the decay happened or not, and whether, before the box is opened to verify which of the two, the cat is in some strange superposed limbo state between dead and alive. This is what has captured the imagination of many although, unfortunately, much nonsense has been said about the matter.

Insofar as the cat is regarded as having only two base states, that question never arises, dead and alive being the only two possible states a cat can be observed in. That is what it means to say a two-level system and we must gloss over the obvious fact that any cat is a many-particle object with an enormous number of states and with no realistic chance of maintaining all the phases and phase relations between them. But, going back to the nucleus itself, or, equivalently, to an excited state of an atom, a quantum system like this can be in a superposition but the decay can only be said to have happened, be complete, when the emitted photon or the products of the decay separate to infinity, with no possibility of re-absorption. The photon has to escape the box and would be so observed even without opening the box, that signalling the decay and the cat's death. With an enclosed box of perfectly reflecting walls, the photon would be reflected back to be re-absorbed by the ground state to return to the excited state, the whole repeated so that there is no decay but oscillations between the two situations, atom excited and no photon or atom de-excited with a photon.

Finally, as regards taking square roots, as in QCF = (NOT)$^{1/2}$, it is worth noting how this operation enlarges the domain of interest. In mathematics, starting with the real line that includes positive and negative numbers, the square root for the latter necessitates an enlargement into the complex plane. In physics, there is similarly an expansion into other dimensions, for instance in the famous example of Dirac's construction of relativistic quantum physics. In aiming to take the square root of the energy–momentum relationship of Special Relativity (Sec. 1.2.4), $E = \sqrt{c^2 \vec{p}^2 + m^2 c^4}$, so as to have linear operators for energy and momentum and thus introduce corresponding first-order derivatives in time and space, respectively, for them, Dirac was led to enlarge the system into an internal spinor space, in his case of four dimensions (Sec. 7.3.2). Indeed, for the case of massless particles with $m = 0$, the Pauli–Dirac equation needs an enlargement only into two dimensions, the 2×2 space of Pauli matrices and intrinsic spin. (The more general result of Dirac introduces both spin and anti-particle extensions, a subject we will return to in Chapter 7.) Interestingly, it is that same spin or two-valued aspect of a quantum coin that allows the (NOT)$^{1/2}$ construction.

4.2.2 The Bloch Sphere

A very useful picture of a two-level quantum system or quantum coin or qubit is provided by the 'Bloch[2] sphere'[21]. Any arbitrary pure state is

---

[2] Felix Bloch, 1905–83, Swiss and American. A pioneer in the quantum description of solids, 'Bloch waves' describe electron propagation and electrical and heat conduction.

depicted as a point on a two-sphere, S^2 (an ordinary unit globe in three dimensions), or, more accurately, in terms of the unit vector from the centre of the sphere to that point. Although we noted previously that there is a three-parameter family of states (two complex coefficients with a normalization constraint), one of them is an overall phase, interest often attaching more to the other two, which can be thought of as the two spherical angles (θ, ϕ), corresponding to latitude and longitude on the Earth. (The 'special unitary' group $SU(2)$ that describes the symmetries of a two-level system is thus viewed as a 'bundle' of a base manifold, S^2, and a fibre, $U(1)$, the latter representing the arbitrary phase that can take any value from 0 to ∞.)

Quantum evolution can then be viewed alternatively in terms of changes of this unit vector. Thus, for example, under Hermitian Hamiltonians, pure states evolve into other pure states with a unitary transformation given by the Schrödinger equation. This can be viewed instead as a classical vector rotating (which does not alter the length of the vector) through some angles as described by a corresponding 'Bloch equation'. In nuclear magnetic resonance (nmr) and its various applications, such as magnetic resonance imaging (mri), this geometrical picture of a vector rotated through some angle, realized by some appropriate magnetic field applied for some time duration, has proved very convenient.

Non-unitary evolution, when dissipation and decoherence may be present, shrink the vector into the sphere. Instead of pure quantum states, we now talk in terms of what are called mixed states, described through a density matrix rather than a ket vector. But the Bloch sphere picture continues to be useful.

Figure 4.7 illustrates the Bloch sphere with fibres at each point (θ, ϕ). While a classical coin or bit has the two states that may be identified with the two poles of the sphere, the quantum coin or qubit's states range over the whole surface of the sphere and along the fibres. This is a vastly (multiply infinitely!) larger and richer space, so a qubit has vastly more potential, even if any observation on it collapses to just two antipodal points. Thus, with 100 such qubits in a quantum computer,

He was an independent co-discoverer of nuclear magnetic resonance, and he derived the 'Bloch equation' for describing the time evolution of the magnetic moment of a charged particle with spin, such as an electron or proton.

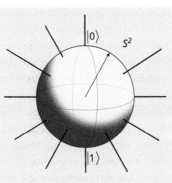

Figure 4.7 The Bloch sphere. States of a quantum spin or other two-level system can be geometrically viewed as points on a two-sphere, S^2, that is, an ordinary globe, together with points on spikes at each point on the sphere. The poles represent the base states $|0\rangle$ and $|1\rangle$; any other point on the surface is a superposition state. Values along the spike from zero to infinity represent a phase that, together with the latitude and longitude location of the spike, provide the three parameters characterizing an arbitrary unitary transformation for a quantum spin or qubit [22].

while the input and output of any calculation are with each qubit in one of its two basis states, during the process of computation, when no observations are being made, many more possibilities can be explored.

As one example, these possibilities allow for an enormous number of primes to be tested to divide into a large composite number, which would be impossible with a classical computer even of many millions of bits. Herein one sees the potential for processes such as factoring a large composite number into two big prime factors that so excited the physics community starting about 15 years ago. Such a factorization being one of the basic principles behind classical cryptography, a quantum computer could have drastic implications for secure transactions in our everyday world [21].

While it is customary to view the two states of a quantum coin as the usually represented north and south poles on the sphere in Figure 4.7, that is but one representation, referred to as the choice of quantization axis along the vertical z-axis. But, as with any quantum system, alternative representations or orientations of the quantization axis are equally valid (Sec. 2.2), so that the base states $|0\rangle$ and $|1\rangle$ can be any two antipodal points on the Bloch sphere. The very spherical symmetry of the sphere puts them all on a par, and an arbitrary superposition state of a quantum spin can be seen as the combination, through Eq. (4.1), of any such pair. The unitary transformations that transform from one representation to another (Sec. 2.2) are rotations from one diameter to another of the Bloch sphere.

Experimental observation of a charged particle with spin is also viewed in this manner in terms of what is called a Stern[3]–Gerlach[4] apparatus, which consists of an inhomogeneous magnetic field. The magnetic moment of such a particle couples to the magnetic field, leading to deflection in opposite directions for the two base states along the axis of the Stern–Gerlach set-up. For any setting of that apparatus, along the z-axis or some other, an entering beam of spin-1/2 particles is always split into just two beams. Both preparation and detection of the two base states are done in this manner.

4.2.3 Pairs of Qubits and Entanglement

In Sec. 4.2.1, we saw that a single quantum coin flip in Figure 4.6 is essentially equivalent to a classical coin in Figure 4.5 but two successive quantum coin flips amount to something different, a switch or logical NOT operation. We consider next pairs of qubits simultaneously. Such systems are the basis of other logic gates, such as exclusive-OR (XOR) or controlled-NOT (CNOT) and, indeed, all logic gates required for computing can be built out of such a pair.

Using the language of quantum-mechanical spin, of two states, up/down ($|\uparrow\rangle$ and $|\downarrow\rangle$), for each qubit (Figure 4.3), states of the pair can be viewed by placing two of them in a Dirac ket (or corresponding bra). Thus, there are four states in all for the pair, which may be denoted ($|\uparrow\uparrow\rangle$, $|\uparrow\downarrow\rangle$, $|\downarrow\uparrow\rangle$, $|\downarrow\downarrow\rangle$), as shown in Figure 4.8. Each ket represents a product of the states of the two qubits and thus this basis set for the pair is said to be 'separable', the wave function factorizing into a product of functions of each qubit.

As in all quantum systems, however, this is but one representation for the pair. Any other linearly independent set of four states can also

---

[3] Otto Stern, 1888–1969, German and American. Experimental physicist, one of the fathers of molecular beam techniques and a co-discoverer of the proton's magnetic moment.

[4] Walter Gerlach, 1889–1979, German. Co-discoverer of spin magnetic moment.

$$| \uparrow\uparrow \rangle$$
$$| \uparrow\downarrow \rangle$$
$$| \downarrow\uparrow \rangle$$
$$| \downarrow\downarrow \rangle$$

Figure 4.8 The four states of two qubits in a separable representation, each ket describing the product of the up and down states of the two spins.

be a valid representation of the pair system. Indeed, invoking again the general principle of linearity and thus the superposition principle but now of two-qubit states, the four linear combinations of the set that are shown in Figure 4.9 are just as good a basis for describing all the physics of the pair. These states differ, however, in a fundamental respect in not being separable. None of them can be factorized into a product of states of single qubits. They are said to be 'entangled', another concept central to quantum physics when one goes beyond a single particle or degree of freedom.

From its very first definition by the founding fathers of the field, this property has been seen as central to quantum physics and one responsible for much of its non-intuitiveness from a classical perspective. The particular set in Figure 4.9 is distinguished among all other (infinitely many) possible representations for a pair of qubits by being at the other extreme of separability from the one in Figure 4.8, being 'maximally entangled'. They are named the 'Bell[5] states' or 'Bell basis', for

$$|P^{\pm}\rangle = (| \uparrow\uparrow\rangle \pm | \downarrow\downarrow\rangle)/\sqrt{2}$$
$$|S^{\pm}\rangle = (| \uparrow\downarrow\rangle \pm | \downarrow\uparrow\rangle)/\sqrt{2}$$

Figure 4.9 The four Bell states of two qubits. In contrast to Figure 4.8, these are now no longer separable but entangled in not being decomposable as products of the individual spins. Indeed, they are maximally entangled and provide an alternative basis to the set in Figure 4.8 for describing a general two-qubit state.

---

[5] John Stewart Bell, 1928–90, Northern Irish. Theoretical physicist who worked on accelerator design and on the foundations of quantum physics. He is credited with having brought the question of quantum interpretation and items such as non-locality and realism from merely philosophical and semantic discussion into the realm of testable experimental physics. He is viewed as a founding father of the field of quantum information.

another pioneer in our understanding of this basic concept of quantum physics [21].

The two states, named $S^{\pm}$, have definite values of total spin angular momentum, $S = 1, 0$, and definite projection $S_z = 0$ on the z-axis (the other two states do not have definite values for these quantities). However, there are no definite values for the spin projection of either spin; the only definite feature is that when one is up, the other is necessarily down. This kind of correlation brings out an important aspect of quantum physics, that the combined system may have definite properties of some physical quantity even when the individual sub-systems do not. The operators of total spin and individual spins do not commute and cannot be simultaneously defined (Sec. 1.2.5). In this, they are analogous to classical concepts such as the 'saltiness' of NaCl, which does not reside in either of the component elements (are both indeed poisonous!), the concept being relevant only to the compound. While saltiness is at least a property that lies outside physics and chemistry, entanglement in quantum physics strikes one as even more unusual because the property of spin projection is an attribute applicable to either the full system or a sub-system. Yet the combined system may have a definite value while the individual sub-systems do not. This is the essence of quantum entanglement.

Just as a qubit represents any two-level system in physics, a pair of qubits can be a stand-in for a four-level system in any branch of physics. In a matrix representation, states would now be described by a column vector of four entries (this for a ket, the corresponding conjugate row vector for the bra) and operators by 4×4 matrices. Besides the unit matrix, there are 15 linearly independent matrices (again with many alternative representations), the counterpart of the three 2×2 Pauli matrices for a single qubit. The symmetry group is now called $SU(4)$, a higher-dimensional analogue of $SU(2)$ for a qubit.

Using the separable basis to describe the four vectors, with one non-zero unit entry in each of the four possible positions, an operation such as CNOT can be described by the matrix shown in Figure 4.10. Its action on the four basis vectors is to leave the first two unchanged but interchange the second two; that is, depending on whether the first spin is up or down, the other (second) spin is either left unchanged or flipped, respectively. The state of the first qubit is said to 'control' the action that takes place on the second, 'target' qubit while itself being left undisturbed. Similarly, any 4×4 unitary matrix can be interpreted in physical

$$\begin{pmatrix} |\uparrow\uparrow\rangle \\ |\downarrow\downarrow\rangle \\ |\downarrow\uparrow\rangle \\ |\uparrow\downarrow\rangle \end{pmatrix} = \begin{pmatrix} 1 & 0 & 0 & 0 \\ 0 & 0 & 0 & 1 \\ 0 & 0 & 1 & 0 \\ 0 & 1 & 0 & 0 \end{pmatrix} \begin{pmatrix} |\uparrow\uparrow\rangle \\ |\uparrow\downarrow\rangle \\ |\downarrow\uparrow\rangle \\ |\downarrow\downarrow\rangle \end{pmatrix}$$

Figure 4.10 The 4x4 matrix for the CNOT, or controlled-NOT operation. When the second, 'control', spin is up, the first remains unchanged, but when the control is down, the state of the first is flipped between up and down.

terms in its action on the pair, as seen in any representation. This CNOT operation is also sometimes called XOR (exclusive-OR).

4.2.4 Quantum Teleportation

It was noted in Sec. 4.2.1 that quantum physics forbids certain operations such as cloning or duplicating an arbitrary state. Special-purpose machines that duplicate specific states are possible but what is not is the duplication of a general superposition of orthogonal states, a general-purpose cloning of quantum states. But the same linearity that forbids such cloning allows a procedure for transporting an arbitrary state from one location to another without physically moving the system. Such 'quantum teleportation' is again a nice illustration of basic features of quantum physics.

Consider, for instance, two parties, A and B, that hold two ends of an entangled state of two qubits, also denoted A and B, say the fourth Bell state in Figure 4.9, which is called the 'singlet state'. This name originates from use in atomic and nuclear physics when two spin-1/2 particles combine into a total angular momentum of zero, $S = 0$, for instance in the ground ^1S state of the two-electron atom helium (see Sec. 2.2.1). There is only one such combination and state. Such a zero angular momentum has, of course, also zero projection on any axis and, in particular, $S_z = 0$. When one spin is up with respect to that axis, the other perforce is down. But which one is up or down does not matter for two identical particles and, therefore, the physical eigenstate is the linear combination that is the fourth Bell state. Its companion, the third Bell state in Figure 4.9, also has $S_z = 0$ but it is a 'triplet' with $S = 1$, while the other two Bell states do not have a definite value of S_z but are linear superpositions of ± 1.

Given such an entangled singlet state (AB) whose two parties and two sub-system qubits, A and B, have separated, perhaps even to large distances, we wish to teleport from A to B the state of a third qubit, C, that is in the general superposition described by Eq. (4.1). A cannot measure C and determine the values c_i, any measurement giving either of the two base states with probabilities $|c_i|^2$, the complex amplitudes themselves out of reach. But what A can do is to couple (in a quantum way that preserves phases) its end of the entangled state to C, and form a three-qubit state, ACB; it is then possible to perform a Bell measurement on (AC) at that end. That is, a joint measurement of the pair is made in terms of the base states of Figure 4.9. The four Bell states of (AC) forming a complete set, the product (AB)C can always be re-written as an expansion in terms of them, with coefficients representing the other qubit of B (the simple algebra involved is not shown here). These coefficients will involve c_i, whose values still are unknown but with A able to see what combinations of the two are involved in each coefficient. Thus, not surprisingly, for the singlet Bell state (AC), B's qubit will be seen to be in the same superposition as in Eq. (4.1). For the other three Bell states, there will be differences in the signs of c_i or of the spin arrows. A can then tell B through a classical channel (perhaps a telephone) what operation needs to be performed at that end, typically flipping the spin or multiplying by a minus sign, to put B's qubit into the same form as in Eq. (4.1). Of course, if A measures a singlet in the Bell measurement, the message will be for B to do nothing. The net result is that the qubit C's state in Eq. (4.1) appears at B's end as the state of that qubit, B. The 'state' has been teleported from A to B.

It is worth emphasizing what exactly is involved in the above procedure. It is only the state that is teleported, not any physical entity such as qubit C. B and C may even be entirely different two-level systems; for example, one may be an electron, the other a proton or a macroscopic object such as a Josephson[6] junction (or even a cat!) which can be in one of two configurations. It is only the state of C that appears

---

[6] Brian David Josephson, 1940, Welsh. Theoretical physicist known for contributions already as a student and especially for discovering a fundamental phenomenon of superconductivity, that quantum-mechanical tunnelling leads to current flow and oscillations between superconductors separated by a barrier of normal matter. This effect has become the basis for precision measurements of magnetic fields and of the fundamental constant $e/\hbar$. He later turned his attention to biology, transcendental meditation, and mind–body problems.

as the state that B is put into. Next, the entangled state between A and B is not enough: some classical information also has to be transferred from A to B to achieve the teleportation. Thus, there is no violation of Einstein's Special Theory of Relativity, as the classical information transfer involved ensures that the transfer is restricted to being below light speed.

Strikingly, A manages to teleport C and the information content in Eq. (4.1) without knowing the values of c_i; indeed, it is imperative that A is ignorant of those coefficients, all that information being lost at the moment when A makes the Bell measurement (A knows only which of the four Bell states of (AC) is realized without knowing anything about C's state - or A's!), the same moment the information appears in B's hand. There is thus no violation of the no-cloning theorem, even across a remote separation. There is only one copy of the state in Eq. (4.1), first as qubit C's state at A's end and later as qubit B's state. A ends up, of course, entangled with C while disentangled from B. All these are necessary consequences of the linearity of quantum physics, while illustrating its internal consistency.

A very similar discussion applies to another important application in quantum information, namely secure key distribution between A and B to establish quantum cryptography. Again, A and B share both a classical and a quantum-entangled channel. That any disturbance of the latter by a third party eavesdropper will be manifest to A and B underlies the security of their exchange. All instances when they detect such disturbance upon comparing notes later through a classical channel can be simply discarded and they can proceed with confidence using the undisturbed exchanges to establish the desired key distribution.

4.3 Qubitcoins

We live in times when the bitcoin has just started 'circulating'. Currencies and coins evolved for keeping track of transactions between persons of commodities or services. From antecedents in uniform cowrie shells, certain plant seeds, and stone wheels, we finally settled on precious metals and coinage, as in Figure 4.1. It is only 50 years ago that this 'gold standard' was jettisoned and the currency's worth became based on the strength of the economy of a nation-state. Side by side, keeping track of transactions has evolved from markings on clay tablets to numbers on transaction sheets and bank statements. But, today, the hundreds of billions of dollars traded in a day exist only in electronic cyberspace. A completely online cyber currency, decoupled from any nation, such as the bitcoin, is part of this trend. As in the physics of motion from bat to fielder (Sec. 8.5), what counts is the transaction between A and B, the paths in between (Figure 7.1) being inessential, however convoluted they may be through multiple banks. The discussion in this chapter of quantum coins may leave us speculating on the even more, literally, mind-boggling future of a 'qubitcoin' world ahead.

5

Symmetry

5.1 Symmetries Around Us

The concept (or principle or metaphor) of symmetry pervades our language and our physics. As one of the animal kingdom's bilaterally symmetric creatures, it is an inescapable observation from an early age that the left and right half of our bodies are similar. (This is from the outside, the inside being very unsymmetrical.) Indeed, they are mirror symmetric in terms of a mirror plane down the middle that reflects one side to the other. Such a mirror or reflection symmetry, or 'parity symmetry' in physics usage, is one of the simplest examples of symmetries.

Elsewhere, we see in both the animate and inanimate world around us other types of symmetries, such as 3-, 4-, or n-fold symmetries. Petals of many flowers, or sea stars display this, a rotation through $2\pi/n$ (a full rotation through 360 degrees is named 2π) restoring the object's appearance (Figure 5.1). Indeed, this is the proper way to describe a symmetry, in terms of some transformation such as reflection or rotation that takes the object into itself, that is, leaves the object unchanged. Note that reflection is different from rotation through π (through 180 degrees), which is also a two-fold symmetry, two such transformations returning to the original configuration. Such a rotation, however, does not take a right hand into the left hand, so that reflection is a distinctly different symmetry.

The inanimate world also has symmetries, as seen in many crystals. The standard example is of snowflakes, which all have a six-fold or hexagonal symmetry (Figure 5.2). Both salt and sugar crystallize in cubic symmetry, while many minerals form hexagonal or octagonal shapes, again symmetrical objects that under various transformations go back into themselves. Some viruses exhibit great visual beauty in their symmetries with very large n (Figure 5.3). There is, finally, the 'perfect' symmetry of a circle or sphere that may be seen as the $n \rightarrow \infty$ limit of such rotations, these objects looking the same with respect to any

Figure 5.1 Flowers with 13-fold and many-fold symmetry, and a starfish with five-fold symmetry. Sarah Cowell <http://www.beeginnerbeekeeper.com/single-and-double-flowers>; Alain Feulvarch <https://en.wikipedia.org/wiki/Starfish>.

Figure 5.2 Six-fold symmetry of snowflakes. Kenneth Libbrecht <http://www.its.caltech.edu/ atomic/snowcrystals/photos/photos.htm>.

diameter, that is, under rotation through an infinitesimally small angle. Note how ordinary language uses this as a metaphor when it describes 'a well rounded argument'.

The immediate visual symmetries of rotation and reflection extend to more sophisticated ones but again with the feature of some transformation and an associated symmetry of interest. (Transformations and symmetries are inseparable so that there is considerable overlap between this chapter and Chapter 2.) Thus, palindromes have a

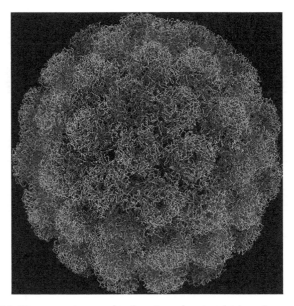

Figure 5.3 Electron micrograph of a virus showing high-order symmetry. Phoebus87 <http://en.wikipedia.org/wiki/File:Symian_virus.png>.

fascination from early childhood; whether in numbers or phrases, they read the same forwards and backwards. (A familiar example is 'Madam, I'm Adam'.) This is with respect to a transformation reversing the direction of flow. Music plays on such ascending and descending scales that come together as aesthetically pleasing to our ears. A musical palindrome by Bach[1] is shown in Figure 5.4.

In physics, it was natural from the very beginning to consider such 'time reversal' symmetry, Galilean and Newtonian mechanics having a symmetry under this transformation. Those laws of motion, 'the laws of physics', are symmetric under such time reflection, and this has profound consequences throughout physics. A major philosophical question is how to reconcile such a symmetry of the laws that are symmetric under a change of sign of time down at the microscopic level with what

---

[1] Johann Sebastian Bach, 1685–1750, German. Organist, harpsichordist, and composer, one of the fathers of Western classical music. He is revered by all musicians and composers, great and small, who followed him over the centuries, and his musical exercises are still taught in the first lessons to music students. He is as central a figure in Western classical music as Newton or Einstein is in physics.

Figure 5.4 A musical palindrome, by Johann Sebastian Bach. <http://www.gfsmaths.com/crab-common.html>.

appears to be a unidirectional flow of time in our macroscopic everyday world. The same question of time reversal symmetry, or 'invariance' with respect to the transformation $t \rightarrow -t$, also dominates quantum physics. Indeed, except for a few limited decays of unstable elementary particles, recently demonstrated [23] directly and unambiguously for 'B mesons', all other interactions 'respect' this symmetry.

5.1.1 Symmetries in Mathematics

Symmetry is also important in mathematics, from the very elementary to the most advanced, and use of symmetry considerations is a powerful tool in a mathematician's or physicist's toolkit. As a first example related to the parity symmetry mentioned at the start of the chapter, even or odd distribution around some mean provides a simple illustration. In summing (or integrating if a continuously distributed function $f(x)$) such numbers, for an even distribution one can simplify the work by a factor of two by considering the sum of just one side and doubling it. For an odd distribution, even further simplification attaches, in that the sum is clearly ('by symmetry') zero without having to do any further computation, values on one side cancelling with the symmetrically equal but opposite contributions from the other side.

Extending further, say to a function of two variables $f(x, y)$, it is immediately apparent that there is a difference between a function such as

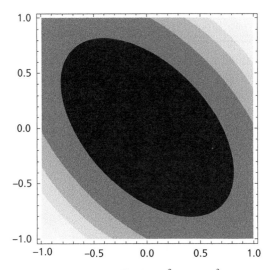

Figure 5.5 Symmetry contours of $f(x, y) = x^2 + xy + y^2$ as concentric ellipses on 45-degree axes.

$(x^2 + y)$ and $(x^2 + xy + y^2)$. Under a transformation that interchanges x and y, the latter is unchanged while the former is not. The second function is distinguished in having this interchange symmetry, a little less obvious perhaps than previous examples but nevertheless also a transformation and an associated symmetry. Geometrically, contours of the two functions display this symmetry, the latter's being ellipses around the axes $(x \pm y)$ tilted at 45 degrees with respect to the horizontal and vertical (Figure 5.5).

The example considered in Sec. 1.2.1, of simplifying the definite integral $\int_{-\infty}^{\infty} \exp(-x^2)dx$, was similar. In that section, this integral was viewed under the theme of adjoining an extra dimension but it also illustrated a symmetry aspect that lay behind the simplification. By adjoining $(-y^2)$ in the exponent, the resulting function, $(x^2 + y^2)$, was even more symmetrical than the second of the functions in the previous paragraph, describing now a circle rather than an ellipse. Thus the product of two somewhat more difficult Gaussian integrals in x and y, when viewed with the circular symmetry of two dimensions, reduced the integral over angle ϕ to the trivial 2π, while the other, radial, integral also simplified into an exponential integral. The two-dimensional integral with circular symmetry is simpler than the one-dimensional linear one with parity symmetry.

The above is a simple but characteristic example of exploiting symmetry to simplify calculations. This problem also brings out another feature, that in the presence of symmetry, the same problem can be viewed from different perspectives, coordinate systems in this case. The double integral separates into a product of two in either Cartesian or circular coordinates. The former view gives the Gaussian integral (rather, a square of it), while the latter is 2π times the exponential integral. This feature becomes even more prominent in quantum physics, as we will see.

Another example appeared in Sec. 1.2.3, where Lagrange multipliers were used to estimate the maxima of a function xyz subjected to the constraint $x^2 + y^2 + z^2 = R^2$; this similarly affords an illustration of the power of symmetry arguments. With both the function and constraint completely symmetric under interchanges of the three variables, symmetry would demand that any solution we are seeking will also have that property and thus $x^2 = y^2 = z^2$, leading immediately to the result in Sec. 1.2.3, that the maxima occur at coordinate value $R/\sqrt{3}$, with no other detailed considerations necessary!

Turning to geometry, yet another example of transformations is provided by projective geometry, wherein points and lines of ordinary Euclidean geometry are viewed instead on a common footing, with a complete duality in interchanging them. The Euclidean geometry studied in secondary schools does not have such duality, but in projective geometry any theorem involving points and lines remains true under such an interchange. There is thus a symmetry between points and lines. A famous diagram of projective geometry, attributed to Desargues[2], is presented in Figure 5.6. It displays 10 points laid out on 10 lines, satisfying the rule that each point lies on three lines and, dually, each line runs through three points. As per its name, projective geometry plays a central role in perspective in art and architecture. Figure 5.6 may be viewed in two ways. From point P, lines are drawn to connect

---

[2] Gerard Desargues, 1591–1661, French. Architect and engineer, one of the founders of projective geometry.

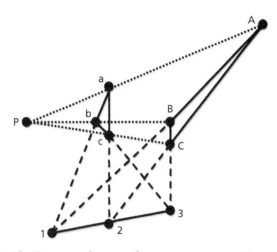

Figure 5.6 The Desargues diagram of projective geometry. Ten points lie on 10 lines, with the incidence relation that every line contains three points and every point lies on three lines. The two triangles, abc and ABC, have like vertices connected by rays from point P, while like sides, when extended, intersect on a common line, 123. The triangles are said to be 'in perspective' with respect to point P and line 123. Note the duality between points and lines that is characteristic of projective geometry. The triangles abc and ABC may lie either in the same plane or not, so that the diagram works equally well as a planar two-dimensional diagram or in three-dimensional space [24].

like vertices of the two triangles abc and ABC. On the other hand, like edges of those triangles intersect, when extended, at three points that, remarkably, lie on a common line. Therefore, it can be said that the two triangles are 'in perspective' with respect to point P and line 123.

A further remarkable aspect is that Figure 5.6 works equally well as a planar diagram or in three dimensions, each of the two planes abc and ABC then oriented generally in space. When the planes are parallel, the three points 1, 2, and 3 recede to infinity, as does their line, to become the line at infinity. An important distinction between Euclidean and projective geometry, again a central feature in perspective, is the removal of any distinction between points at infinity and those at finite locations.

Also important is the idea of more or increasing symmetries shown by the functions $x^2 + y$, $x^2 + xy + y^2$, and $x^2 + y^2$. The second has higher

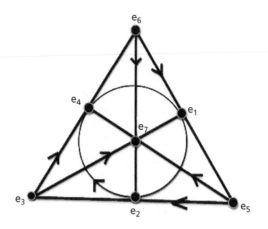

Figure 5.7 The Fano plane. Similar to Figure 5.6 but now for seven points and seven lines (the inscribed circle is counted as on a par with the others, no distinction being made in projective geometry), with the added feature that any pair of points lies on a unique line, making this a projective plane called PG(2,2). This same diagram with added arrows, as shown, represents the multiplication table for octonions, the e_i being the seven independent square roots of (-1), the product of any two giving the third on that line, with a ± 1 factor when along/against the arrow direction [24].

symmetry under x and y interchange than the first (no symmetry at all) and the third even higher (actually perfect symmetry as of a circle over an ellipse). Similarly, Figure 5.6 has a counterpart in projective geometry shown in Figure 5.7, in which seven points and seven lines display the same feature that each point lies on (is said to be incident to) three lines and each line is incident to three points but can be said to have an even higher symmetry. This focus on incidence relations of just the points shown makes these diagrams work also in finite geometries with just those 7 or 10 points, the rest of the points being shown to display the lines as continuous, though not really relevant to the discussion. This is a geometry of just a few finite number of points. For this reason, the inscribed circle in Figure 5.7 that connects three points, albeit at infinity, is just as legitimate a line as any of the edges or medians of the triangle. Figure 5.7 has even more symmetry than Figure 5.6 in having yet another feature, that every pair of points lies on a distinct line, which is not true for Figure 5.6, where, for instance, point P has no lines joining it to points 1, 2, and 3. Indeed, Figure 5.7, called the 'Fano[3] plane', is also important in projective geometry and is the simplest 'projective plane' [25].

The Fano plane in Figure 5.7 plays a role also in a branch of mathematics called Design Theory and is related to so-called symmetric designs where two sets of quantities are placed into incidence classes with respect to each other. These may be anything — say, varieties of some agricultural product such as potatoes and varieties of fertilizers applied to them. In choosing to measure the effectiveness of the latter to improve yields of the former, and therefore designing fields so that each potato variety sees each of the fertilizers, such symmetric designs are used, pointing to the usefulness of this mathematics in many applications of experimental design [24, 25].

Curiously, the same diagram also appears in an entirely different mathematical context, namely in describing the only two consistent arithmetics (more technically, division algebras) beyond real and complex numbers. Most people know of real and complex numbers (based on i, 'the' square root of -1), that they can be multiplied and divided. Mathematicians and physicists also deal with two other number systems, called quaternions, which employ three independent square roots of -1, designated (i, j, k), with $i^2 = j^2 = k^2 = -1$, and octonions, which similarly use seven independent square roots. There are no other consistent systems, these four exhausting the possible division algebras. A set of rules has to be prescribed for the multiplication of any two different square roots in terms of a third to close the algebra, and these rules can be most conveniently kept track of through Figure 5.7. Together with arrows drawn on each of the seven lines, any of the seven triplet lines of the diagram can designate these rules for quaternions, although it is most common to use the circle. With the quaternions' (i, j, k) as the three points on this circle in cyclic (clockwise) order with a clockwise arrow on that line, we set $ij = k, jk = i, ki = j$, whereas in the reverse order against the direction of the arrow a minus sign is attached, and we set $ji = -k$. A similar but slightly more complicated prescription [26] works for the seven square roots of -1 of octonions placed at the seven points of the diagram in Figure 5.7 and using all seven lines shown.

---

[3] Gino Fano, 1871–1952, Italian. Mathematician with contributions to geometry and group theory.

5.1.2 Symmetries in Physics and Conservation Laws

An important consequence of symmetry in the equations of motion in physics is that they imply a corresponding constant of the motion or a conservation law, that a corresponding quantity does not change during the motion. The famous laws of conservation of energy, linear, and angular momentum are associated with such symmetries. These absolute conservation laws being some of the most fundamental entities of physics, the consideration of symmetry is therefore central to the subject. Emmy Noether's[4] discovery of the link between symmetries and conservation laws is among the most important theorems of mathematical physics.

There is already in Newton's third law of motion a symmetry, that the force exerted by a particle (1) on another particle (2) is equal and opposite to that exerted by 2 on 1. This holds true for any force, whatever its origin, including forces of electricity and magnetism that were not known in Newton's day, or the forces of strong and weak interactions between elementary particles and nuclei that were discovered even later. Colloquially rendered as 'every action has an equal and opposite reaction' and applied widely (if not always accurately!) even outside of physics, this is one of the powerful and familiar metaphors of our language, one with its origins in physics. With all forces between pairs cancelling out, the total system has no net force on it from the mutual interactions contained within, and thereby the total linear momentum, denoted $\vec{p}$, is conserved. If all the torques due to these internal pairwise forces also add to zero, the total angular momentum, denoted $\vec{\ell}$, is also conserved.

Later, in the Lagrangian and Hamiltonian formulation of Newtonian mechanics, this result takes a slightly different form, that the Lagrangian depends only on internal separations between the particles in a N-body system, not on the centre of mass coordinate, with the result that the derivative of the Lagrangian with respect to the corresponding

---

[4] Emmy Noether, 1882–1935, German. Mathematician with contributions to abstract algebra and physics, where she is best known for establishing the connection between symmetry principles and conservation laws. This 'Noether's theorem' has been described as 'the most important mathematical theorem ever proved'. Notwithstanding prejudices against women in academia, she joined the mathematics school at Göttingen under the invitation of Hilbert and Klein. Many major physicists described her as the most important woman in the history of mathematics.

velocity is conserved (this derivative is termed the conjugate momentum). This is the centre of mass momentum or, for an angular coordinate, the angular momentum, these quantities then not changing during the motion. An elegant expression of this is that if the Lagrangian does not depend on such a coordinate, it is invariant with respect to additive changes, so that when such additions due to translations in space or rotational orientation do not affect the Lagrangian, the corresponding conjugate momenta are conserved.

Most often, the Lagrangian and Hamiltonian are, respectively, the difference and sum of the kinetic and potential energy of the system. Instead of vector forces and torques, it proves much more convenient to handle these scalar energies. The further feature that when a coordinate does not enter into the Lagrangian, that is, there is a symmetry with respect to it, the equations lead to a conserved quantity makes this formulation especially appealing and convenient. Not surprisingly, most of physics, whether classical or quantum physics or quantum field theories, uses Lagrangians and Hamiltonians.

These considerations apply not just to a spatial coordinate, whether linear or circular, but equally to time. If the Lagrangian does not involve explicitly the time variable, t, that is, is invariant with respect to translations in time, another conservation law applies, the conservation of energy, the quantity conjugate to time, just as momentum, linear or angular, is conjugate to spatial coordinates. A system as a whole, with no external forces that may be turned on and off for some finite time period, has its total energy fixed, whatever internal exchanges may take place between the sub-systems of which it is composed.

The Kepler problem of two masses bound together by gravitational attraction, or a Coulomb pair of oppositely charged electrical particles similarly bound, are both systems with a spherically symmetric potential that varies as $1/r$ (correspondingly, the force is 'inverse square'), that is, depends only on the separation between the particles and not on how that separation is oriented in space. The Lagrangian in spherical polar coordinates not depending on any angles, angular momentum is conserved. Whatever initial value of $\vec{\ell}$ they start with, that value is conserved. This being a vector quantity fixed, therefore, in magnitude and direction, the motion is necessarily confined to a plane orthogonal to that direction. Indeed, any potential, $V(r)$, that is, one that depends only on r and not $\vec{r}$, shares this property. (With no t involvement either, energy is also a constant of the motion, only swapping between kinetic and potential energy during the orbit but the total fixed at whatever initial value was set.) The Kepler–Coulomb problem is special among these, in that there is something additional, namely the orbits closing in ellipses (Sec. 1.2.5).

That a closed orbit means there are even more conserved quantities has been recognized from Newton's times. The eccentricity of the orbit and the direction of, say, the major axis (the minor or any other could also have been chosen) together define a so-called 'Laplace–Runge[5]–Lenz[6]' vector, $\vec{A}$ (Figure 1.12), which is also conserved, along with $\vec{\ell}$. The pointer to a 'higher' symmetry of the Kepler–Coulomb problem (and one other, the harmonic potential r^2) than just the one under three-dimensional rotations that is shared by all spherically symmetric potentials found more complete explanation in quantum physics, as discussed for the hydrogen atom in Sec. 1.2.5. Whenever there is an additional symmetry, it is a pointer to degeneracies in the spectrum (that different physical states share the same energy) and to alternative coordinate systems or groups of commuting operators that can describe the system. The connection between symmetries and conserved or invariant quantities (and degeneracies of spectral levels in quantum systems) is true throughout physics.

The theme of 'broken', especially slightly broken, symmetries again takes full significance in quantum physics, but also can be seen in classical physics with the Kepler–Coulomb problem. Any admixture of some other dependence into $1/r$, however slight and even if itself also spherically symmetric, changes the situation fundamentally. While angular momentum is still conserved and the orbits still lie in the plane orthogonal to it, no additional $\vec{A}$ will exist, as manifest in the fact that the orbits will not close. The famous example is that of the precession of the perihelion of planetary orbits. Since the planet–Sun system also experiences other forces or potentials, for example from the presence of other planets, orbits are not closed, fixed ellipses. Instead, the axes precess. Since these other forces are very weak compared with the dominant attraction of the Sun, the symmetry is broken only slightly

[5] Carl David Tolme Runge, 1856–1927, German. Mathematician and physicist, with contributions in numerical analysis and spectroscopy.

[6] Wilhelm Lenz, 1888–1957, German. Physicist known for his work on the hydrogen atom, the 'Ising model' in statistical physics, and for the training of many prominent physicists at his institute of nuclear physics.

and the precession is small, so that the notion of the ellipse retains meaning and the additional effect is seen as only a slight change in the position of the axes after each orbit. Not surprisingly, the largest planet, Jupiter, has the largest effect on other planetary perihelion precessions.

While all this has been known since Newton's time and was worked out centuries ago, there remained a tiny displacement of 43" of arc per century in the precession of Mercury's axes, which stubbornly persisted until it yielded to Einstein's General Theory of Relativity, which gives a gravitational potential that departs slightly from Newtonian $V(r) = -GMm/r$. Einstein accounted for the discrepancy, a triumphant prediction of his theory of gravity. The atomic counterpart displays something similar, weak forces such as the one coupling electronic spin, $\vec{s}$, and orbital angular momenta, $\vec{\ell}$, being scalar product potentials, $\vec{s} \cdot \vec{\ell}/r^3$. While still spherically symmetric, this changes the strict $1/r$ nature of the potential, with resulting effects on degeneracies and other features of the spectral levels. We will return to this in Sec. 5.2.4.

5.2 Symmetries in Quantum Physics

While the consideration of symmetry has been important in physics from Newtonian times, it took on even more significance from the start in quantum physics. The founding fathers recognized well its importance as a guiding principle in formulating the new mechanics. Weyl[7], Wigner, and Dirac, in particular, have written eloquently about it. Symmetry (in the mathematical sense) in the basic equations has been equated with beauty, just as in the natural world flowers and snowflakes are admired for the symmetry of their beauty (Figures 5.1 and 5.2). We will return to this in Sec. 5.2.5.

With a Lagrangian formulation also natural for quantum mechanics or quantum field theories, the connection between symmetries under translation or rotation in space or translations in time, and

---

[7] Hermann Weyl, 1885–1955, German. Mathematician, philosopher, and physicist, one of the most influential mathematical physicists, who had close associations with Einstein, Schrödinger, Felix Klein, Hilbert, and others. He studied the distribution of eigenvalues of the Laplacian operator, introduced the concept of gauge, the 'Weyl tensor' in Riemannian geometry, the 'vierbein' in General Relativity, and, through many other contributions to group theory and representations, is one of the most important figures in the mathematical formulation of quantum physics and General Relativity. He is also a key figure in mathematical philosophy.

corresponding conservation laws of momentum, angular momentum, and energy remain equally valid in quantum physics. These are among the most fundamental features of physics and are intimately linked to symmetries in space and time. But, in addition, it also became natural to consider other space–time symmetries, such as reflections under space and time, and Lorentz transformations between different inertial frames, given that quantum physics developed soon after physics had imbibed the Special Theory of Relativity, which gives greater emphasis to this larger set of symmetries than does classical physics.

5.2.1 Discrete Symmetries

Consider first the simplest discrete transformation, reflection in space, $\vec{r} \rightarrow -\vec{r}$, called the parity transformation, or just parity for short, P. In Cartesian coordinates, all three coordinates are reversed, ($x \rightarrow -x, y \rightarrow -y, z \rightarrow -z$), while in spherical polar coordinates a point on a sphere is taken to its antipodal one, ($r \rightarrow r, \theta \rightarrow \pi - \theta, \phi \rightarrow \pi + \phi$) (see Figure 5.8). Any quantum system whose potential is invariant under this symmetry transformation (the non-relativistic kinetic energy, involving as it does the squared momentum, is so invariant) will have a spectrum that can be divided into two parity classes, even and odd under this parity transformation. Thus, the free particle in one or

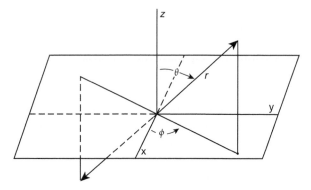

Figure 5.8 Spherical polar coordinates in three dimensions. For any point, the vector $\vec{r}$ is decomposed in terms of its length, r, tilt angle, θ, from the z-axis, and angle ϕ, as shown, measured from the x-axis in the $x-y$ plane to the foot of the perpendicular dropped from the point onto that plane. Reflection of a point through the origin, $\vec{r} \rightarrow -\vec{r}$, is achieved by $\theta \rightarrow \pi - \theta, \phi \rightarrow \pi + \phi$, as can be seen by the corresponding projections shown.

any number of dimensions, the harmonic oscillator, and the hydrogen atom are all examples where energy eigenstates states can be labelled simultaneously with the ± 1 of parity, the Hamiltonian commuting with parity, $[H, P] \equiv HP - PH = 0$, the order in which the two operators are applied being immaterial if $H \to H$ under P. In the spherical representation of the states of the hydrogen atom (Sec. 1.2.5), states are characterized by the quantum numbers (n, ℓ, m) of the operators (H, ℓ^2, ℓ_z), and also have the parity label $(-1)^\ell$.

It is a feature of quantum physics that there is a representation dependence, because simultaneous labelling is only for the set of operators all of which mutually commute, and P does with the other three of this spherical set. But eigenstates in an alternative representation, such as the parabolic for the hydrogen atom (Sec. 1.2.5), whose labels stem from the operator set (H, ℓ_z, A_z), are not simultaneous eigenstates of parity, because P, while commuting with H, does not with the 'polar' vector quantity $\vec{A}$, and thus with its component A_z (it does, however, with the 'axial' vector $\vec{\ell}$). The same is true of the one-dimensional free particle, which among alternative representations, has the 'travelling wave' or 'standing wave' pictures with wave functions $\exp(\pm ipx/\hbar)$ or sines and cosines, respectively. (Note also another dual aspect, of complex or real descriptions, respectively.) The former are simultaneous eigenstates of H and $\vec{p}$, while the latter are of H and P.

Since $\vec{p}$ of momentum and P of parity do not commute, one has to choose between the alternative representations, both complete sets, according to use and context (Sec. 1.2.5). Travelling waves, by definition, move from left to right or vice versa and are, therefore, not invariant under reflection, but rather one wave goes into the other under P. Likewise, the parabolic states of the hydrogen atom, which are superpositions of parity eigenstates (as the exponentials are of sines and cosines), transform under parity into each other, a parabola facing z going into an identical one facing $-z$. With an electric field having a similar transformation under parity, it is natural that the parabolic states are suitable for discussing the hydrogen atom in an external electric field. The field-free atom or an atom in a magnetic field, on the other hand, is more conveniently treated in terms of spherical states. This illustrates the use of symmetry considerations in choosing between alternative representations to work with, a practical aspect of symmetry in the toolkit of a physicist.

A further feature of the hydrogen atom in an electric field is that while all atoms have their internal spherical symmetry reduced to the cylindrical symmetry with respect to the electric field direction, excited states of the hydrogen atom, and they alone (not of any other atom or the ground state of hydrogen), exhibit a further feature that connects to other themes. No other atom has a pure $1/r$ field and its additional symmetry that is linked to the additional degeneracy in the spectrum of states of different ℓ. In particular, this means states of opposite parity are degenerate. An electric field, also being odd (reverses) under this parity transformation, therefore, can mix these states.

With initially degenerate states, even the slightest interaction that mixes them can lead to maximal mixing. Indeed, when there are two degenerate states such as the (in spherical description) $2s$ and $2p$ (with $m = 0$), the mixing is 50:50, and the linear combinations $(2s \pm 2p)/\sqrt{2}$ with such equal mixing are precisely the parabolic states. Although it is straightforward enough for a physics student to see the equal mixing in terms of diagonalizing a 2×2 matrix with equal entries along the diagonal and equal ones along the off-diagonal, it is nice to see symmetry itself as leading to this conclusion without invoking any mathematics. The logic is simple. If the two states are on an equal footing and are mixed, there being nothing to favour one over the other, they are equally probable in the mixture. It is also nice to note another aspect, that this is like going to 45-degree axes in state space, an illustration of the theme in Sec. 2.2.1. The hydrogen atom's Schrödinger equation, which separates in both spherical and parabolic representations for zero field, still does so for the latter in the presence of an electric field.

Consider next reflection of the time coordinate, called time reversal, $T : t \rightarrow -t$. Classical physics, whether expressed through Newton's equations, which are second-order differentials in time or Lagrangians and Hamiltonians, is invariant. The microscopic world is entirely symmetric with respect to reversing the direction of time. A major challenge has always been to reconcile this with the macroscopic world of everyday experience, which clearly is not, but these aspects of time in physics are not of concern here. Here, we consider time reversal in quantum physics. The non-relativistic Schrödinger or relativistic Dirac equation involve only a first derivative in t, but always accompanied by i. Since that also changes sign under time reversal, that operation incorporating complex conjugation, quantum equations of motion are also invariant under this symmetry. Dirac's extension of quantum mechanics to the relativistic domain brought naturally into physics anti-particles, such as the positron, and with it a third discrete transformation called charge conjugation, C, which changes the sign of the charge.

While all interactions known to physics conform to the laws of conservation of momentum, angular momentum, and energy, not all do under the discrete transformations of (P, C, T). Parity, for instance, is a

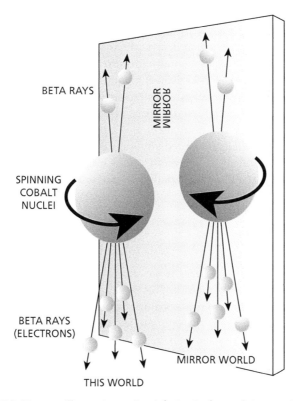

Figure 5.9 Diagram illustrating parity violation in the weak-interaction decay of a cobalt nucleus. The asymmetry in the direction of emission of electrons with respect to the spin of the nucleus violates parity symmetry because the mirror-reflected arrangement would argue for the opposite result. Had weak interactions respected parity symmetry, the two arrangements would have shown the same physics, but parity is indeed maximally violated, as shown first in 1957 in this cobalt decay and other systems. From <http://www.aps.org/publications/apsnews/200112/history.cfm>.

good symmetry of electromagnetic and strong interactions but not of weak nuclear interactions (which cause decays of elementary particles). Indeed, some weak interactions have a maximal violation of parity symmetry. This is expressed by saying that the mirror reflection of such a decay is not observed as a possible physical phenomenon in the laboratory (Figure 5.9). Interestingly, under the combined operations of parity and charge conjugation, that is, under combined CP, the symmetry is restored, the mirror image in Figure 5.9 being indeed a description of the decay of the anti-Co nucleus into the corresponding anti-particles.

On the other hand, T seems to be a good symmetry of all known physics, including weak decays, as in the Co example, except for just a couple of weak interaction decays where a slight, but unambiguous, violation has been experimentally measured [23]. Most interestingly, all quantum field theories have as a good symmetry the combined reversal of all three. That is, for any reaction or decay that we observe, when all particles are reflected in space, replaced by their anti-particles and time reversed, the resultant is also a possible reaction or decay, and identical in all its measured physical values. (The CP operation is, therefore, equivalent to T reversal, and it is indeed CP violation that has been observed in a few weak decays.) So far as we know of current physics, CPT invariance is as absolute a conservation law as those of energy and momentum.

5.2.2 Gauge Symmetries

Quantum physics introduces types of symmetries that had not been considered in classical physics. This is because of the use of complex quantities such as wave functions in quantum mechanics or fields in quantum field theories. The simplest new symmetry is that a complex function can have a change in its phase and if that does not change the Lagrangian and thereby the physics, there will again be a conserved quantity, the corresponding conjugate expressed as the derivative of the Lagrangian with respect to that phase (Sec. 5.1.2). An explanation for the observed law of conservation of (electric) charge was seen as a result of this 'gauge invariance' of the Lagrangian.

Indeed, this went much further, in that, together with a change in the phase of a field such as that of an electron, coupling terms representing its interactions with a vector field must also undergo a transformation to keep the total Lagrangian invariant. This transformation was an already known gauge transformation of the potentials of a classical electromagnetic vector field (Sec. 1.2.4). That the two gauge transformations went hand in hand and that, together with a conserved charge its coupling to an electromagnetic field marked a self-consistent whole, brought a pleasing unity to the subject of charged particles and their interaction with electric and magnetic fields. In a way, starting just with a description of the electron field, its gauge invariance points to the existence of another field, the electromagnetic. A conserved charge of the electron and its coupling to that second field are natural accompaniments, all dictated by the requirements of symmetry. It is one of the beautiful stories of symmetry and its paradigmatic role in physics.

This gauge transformation involving a phase, a single number that can vary continuously from zero to arbitrarily large value, is the simplest and is described as the unitary group, $U(1)$, of just one element, the phase, a scalar number. It can describe electric but also other charges, depending on the structure of the additional field invoked and the coupling to it. Gauge symmetries with higher-dimensional and more complicated groups have also been employed by quantum field theories and it is through such gauge symmetries that modern particle physics handles elementary particles, fields, and their interactions.

5.2.3 Supersymmetry

Operators in quantum mechanics satisfy various commutation relationships. Likewise, in quantum field theory, operators of a bosonic system (spin integer) obey commutation, and those of a fermionic system (spin half-odd integer) anti-commutation relations (Sec. 7.3.3). An even wider symmetry, called supersymmetry, has been invoked with a mix of both aspects. Supersymmetry puts into degenerate multiplets of elementary particles not only bosons and fermions separately but also together. The full set of operators in the system close under a mix of commutators and anti-commutators. Originally invoked for solving some technical problems of field theories, supersymmetry (SUSY) has become widespread, although there is as yet no experimental data in support.

We will consider here an aspect of supersymmetry in quantum mechanics that pertains to many simple systems, both non-relativistic and relativistic. A characteristic of SUSY in field theories is a spectrum, as shown in Figure 5.10, namely a non-degenerate ground state of zero energy identified with the vacuum (with zero value for all quantum

numbers, necessarily a boson) and every other state doubly degenerate
as a boson–fermion pair. Operators that transform among these states
commute with the Hamiltonian (which is why they are degenerate)
while their anti-commutator gives H itself, making for a closed algebra. There is no evidence that every (or even any) elementary particle
we observe comes with such a degenerate partner of the same mass but
spin differing by $1/2$ (of course, particles such as electrons and protons
have partner anti-particles but they share the same mass and spin, and
are also fermions). Even if the SUSY is broken, and the masses are different, there is no evidence for that, or of a duplication in number of all
elementary particles seen, although searches continue.

   In quantum mechanics, however, spectra with the feature in
Figure 5.10 that may be dubbed SUSYQM occur quite commonly
[27]. The free particle in one dimension is already an example. In
either of the two descriptions of travelling or standing waves, there is
one ground state of zero energy. It has necessarily zero momentum
and even parity. All other states, now continuously distributed in
energy, E (Figure 5.11), and not discretely as in Figure 5.10, are doubly
degenerate, either travelling waves in both directions or even and odd
parity partners. The counterpart free rotor in two dimensions, with
Hamiltonian $H = \ell^2/2I$ of angular momentum ℓ and moment of

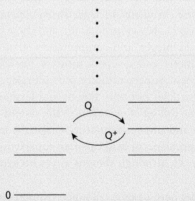

Figure 5.10 A supersymmetric quantum-mechanical (SUSYQM) spectrum. A
non-degenerate ground state at zero energy and all other states doubly degenerate characterize such a spectrum, with operator Q and its adjoint carrying
those pairs into each other [27].

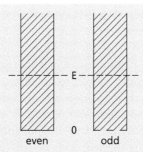

Figure 5.11 The one-dimensional free-particle spectrum as an example of SUSYQM. This continuous spectrum with energy from zero to infinity has a single state exactly at zero energy, all others with non-zero E being pairs in either parity or direction of travel. The absence of $E = 0$ for the odd-parity ladder is indicated by the dashed line at the bottom [27].

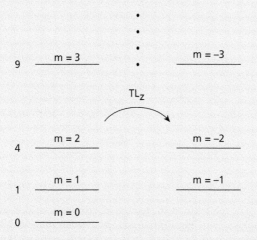

Figure 5.12 The spectrum of a rotor as another example of SUSYQM, now with a discrete spectrum in contrast to the one in Figure 5.11. The operator, Q, which carries a state into its pair partner, is a product of time reversal and angular momentum [27].

inertia I, has a discrete spectrum with eigenvalues $m\hbar$, $m = 0, \pm1, \pm2, \ldots$ for angular momentum and for energy $E = m^2\hbar^2/2I$ (Figure 5.12), returning to the pattern in Figure 5.10.

A relativistic example is of the Dirac electron in a uniform magnetic field. There is now a doubling of that figure, in that for both electron and positron, themselves degenerate, such a spectral distribution applies. Spatial quantization in the magnetic field gives 'Landau[8] levels' that are equally spaced with the cyclotron energy separation and a ground state at a zero-point value of half that energy. But the spin of the electron (positron) also couples to the magnetic field and, with the g-factor of 2 in Dirac theory, this is exactly sufficient to cancel that zero-point value to give a net value of zero for the ground state with Landau quantum number zero and the spin anti-parallel to the magnetic field. Since flipping the spin costs exactly the same energy as the Landau spacing, again because $g = 2$, we have a SUSYQM spectrum of equally spaced eigenvalues starting from a non-degenerate zero-energy state to pairs for all other integer multiples of the cyclotron energy (Figure 5.13) [28].

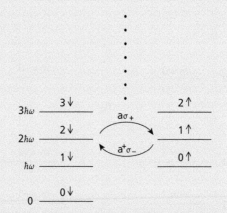

Figure 5.13 The spectrum of an electron in a magnetic field, B, as evidence of SUSYQM in this system. Equally spaced Landau levels along with the coupling of the electron spin to B (with the g-factor exactly equal to 2) lie as shown, with the lowest state at zero energy having Landau quantum number 0 and spin anti-parallel to the field direction. Each increasing n and spin flip costs the same energy, $\hbar\omega$, making all excited levels doubly degenerate. The operator, Q, of Figure 5.10, which transforms between them is a product of the operator, a, that steps down in the ladder of levels, and σ_+, which flips the spin from down to up.

[8] Lev Davidovich Landau, 1908–68, Russian. An outstanding theoretical physicist with many contributions in quantum physics: the density matrix method, superfluidity and superconductivity, phase transitions, plasma physics, quantum electrodynamics, and neutrino physics. He was greatly influenced by Bohr. He developed a great school

5.2.4 Broken Symmetry

The concept of a symmetry that is broken, often only slightly, has already been mentioned in a classical example when the adding of any small perturbation with a different r dependence to the $1/r$ potential leads to orbits that do not close into ellipses (Sec. 1.2.5 and 5.1.2). This is also widely true in quantum systems. Thus, the coupling between the magnetic fields of spin and orbital angular momentum of an electron in the hydrogen (or any other) atom gives rise to an additive potential $\vec{s} \cdot \vec{\ell}/r^3$. This lifts the degeneracy in ℓ of hydrogenic states with the potential now no longer $1/r$, expectation values of $1/r^3$ depending on both n and ℓ, unlike that of $1/r$, which depends only on n (Sec. 5.1.2). However, this contribution being small on the scale of the Bohr energy, the splittings are small so that it still makes sense to view the system as starting from the degenerate limit; indeed, calculations proceeding in this way show the practical importance of symmetry considerations even when the symmetry is not exact.

The SUSYQM example of electrons in a magnetic field in Dirac theory provides another nice illustration. The field theoretic description of electrons and photons, called quantum electrodynamics (QED), leads to corrections to Dirac theory; in particular, the g-factor is slightly different (about 1%) from 2 because of these QED corrections, so that the spectrum in Figure 5.13 is modified, the ground state not exactly at zero (the spin energy not compensating perfectly for the zero-point energy) and the pairs differing also by about 1%.

5.2.5 Spontaneous Breaking of Symmetry

Examples considered so far in classical or quantum systems of broken symmetry are due to some additional, external field that does not have the underlying symmetry of the zero-field Hamiltonian. Thus the hydrogen atom's spherical symmetry stemming from the Coulomb potential may be broken by an applied electric field in some direction which singles out that direction as special. A lower symmetry then

of physics in the Soviet Union, and many prominent physicists were trained by him, in his style of a broad-ranging mathematics and physics training in all fields. The 'Landau–Lifshitz' series of texts in theoretical physics has influenced and educated physicists around the world. He was severely injured in a traffic accident from which he never fully recovered for the last six years of his life.

holds, namely azimuthal (cylindrical) symmetry with respect to that direction, with full spherical symmetry lost. Or, the spin–orbit inter-action considered in Sec. 5.2.4 may maintain spherical symmetry but it lowers the four-dimensional rotation, $O(4)$, symmetry of the pure $1/r$ potential to that of three-dimensional rotations, $O(3)$. As a result, the degeneracy between s and p states no longer applies, is 'lifted'.

One of the most interesting developments, however, in quantum field theories was the realization that, with interacting fields, symmet-ries may be broken internally without having to be introduced from the outside, spontaneously within the system Lagrangian itself. This has proved crucial in formulating the current theory of elementary par-ticles. A main advantage of having a symmetry broken spontaneously is that the theory remains 'renormalizable', and calculations are not plagued by infinities as they are when the breaking is inserted by hand.

A canonical example is the standard model that unified electromag-netic and nuclear weak interactions in a single 'electroweak' inter-action. The two were known from the earliest days to be different in character, the former long range (it and Newton's gravitational inter-action the two famous infinite-range ones) and the latter short range. In physics, there is an inverse relation between the range of interaction and the mass of the quantum carrier: the heavier that mass, the shorter the range (as if the force falls off exponentially, $\exp(-mcr/\hbar)$). The pho-ton being massless (as is also the graviton), electromagnetic forces are of infinite range. On the other hand, weak interactions being confined to about 1 fermi, it was clear that the carrier mass had to be large, ap-proximately 100 times the mass of the proton, so that $\hbar/mc \approx 1$ fermi $(10^{-15}$ m$)$.

But introducing such a mass term through a quadratic potential in the Lagrangian makes the theory unrenormalizable. The standard model's solution to this is that through interactions of vector and sca-lar fields, introduced initially as massless, spontaneous breaking can give one of the vector fields mass. Through such a construction, the vector field responsible for both charge-changing and neutral weak interactions gets massive, the corresponding quanta being the $W^{\pm}$ (ne-cessarily equal in mass because of CPT invariance) and the Z^0, while one other linear combination of the neutral fields remains the massless vector particle that is the photon.

This famous step of unification of forces in physics, the first after Max-well's unification of electric and magnetic forces, rests crucially on the phenomenon of spontaneous symmetry breaking. An even further step is the introduction of the Higgs[9] boson in the theory of elementary particles, which also embraces the strong interactions of quarks and gluons. Again, all particle masses can be attributed to coupling to this Higgs field and, for this purpose, it perforce has to be a scalar (zero spin), so that it couples to all, universally. A recently discovered heavy particle of mass 125 GeV seems to be the quantum of such a Higgs field, which would represent one more step in spontaneous symmetry breaking's key role in the unification of physical forces.

5.2.6 The 'Why' of Symmetry and Its Breaking

Physics, and science more generally, does not usually deal with 'why' questions, only 'what' and 'how'. But it is interesting to ask, given the central role of symmetry in physics, why this should be so. One answer lies in the connection noted between the fundamental laws of conservation and independence from specific frames of reference. Such an independence is of course necessary for the whole enterprise of physics to make sense and for there to be universal laws. If what happened in one laboratory was different from what happened in one down the hall or in another country or, for that matter, on another planet or galaxy, there would hardly be any common physics to discuss. Thus, a translation in space must be a symmetry of the subject. So too, if a measurement today was different from one tomorrow or yesterday, that is, was not invariant under time translations, there would be no validity to the science. Therefore, at least global symmetries such as these are so necessary that it would be hard to imagine physics without them. The corresponding conservation laws of energy and momentum are also, along with some others, such as of electric charge, the ones we see as absolute, with no violations observed.

Indeed, throughout the history of physics, when even some of the formulators seemed to think it necessary to allow violations in building a new mechanics, faith in the absolute validity of these conservation laws was vindicated and often pointed to the correct formulation. This was notably so in the early days of quantum physics, with its seeming statistical aspects that led people to entertain the idea that the

---

[9] Peter Ware Higgs, 1929, British. Theoretical physicist, known for his work on broken symmetry in particle physics. His name has been attached to the quantum of excitation of the associated quantum field.

conservation laws held only in the statistical aggregate but not for each individual event or experiment. However, later and more careful analysis and experiment verified that energy and momentum are conserved in each individual case as well. Our faith in these conservation laws is, therefore, strong; when they seem to be violated, closer examination has always shown either a mistake in theory or experiment or a subtlety not initially appreciated, which upon recognition added further insight into the physics involved.

Other symmetries and related conservation laws, however, are sometimes broken. Parity is a notable example. Also historically interesting because a bias had developed of seeing all space–time symmetries as absolute, as firmly valid as those in the previous paragraph, so that violation of parity invariance, that the total parity on either side of a decay process can differ for weak interactions, came to many physicists as a shock. As an analogy, human (and other animal) faces and bodies in their outward appearance have bilateral symmetry between the left and right halves. Although this is usually slightly violated, many cultures and ideas of aesthetics even seeing beauty in these slight asymmetries, it is nevertheless striking to come across a sculpture such as that shown in Figure 5.14, which has a complete reversal between the two halves of the male/female element. In an interesting analogy to the combination (CP) of parity and charge conjugation, simultaneous reflection about the mid-line, along with interchange of male and female elements, restores the symmetry of the sculpted figure!

Of course, it was soon realized that there was no reason for interactions to display certain symmetries. This is no more than recognizing that while the symmetry of an electron bound to the Coulomb field of a proton is spherical, introducing an electric field in some direction and thus breaking that symmetry will leave the system with lower symmetry, cylindrical with respect to that field direction in this example. In the case of parity invariance, it is intrinsic to the nature of weak interactions that the symmetry does not hold, whereas it does for strong and electromagnetic interactions.

In many cases, the external field being weak relative to internal ones (an atom has internal fields of 10^9 V/cm), the breaking of symmetry is slight, although, as noted in Sec. 5.2.1, there can also be maximal mixing, a maximal breaking of symmetry, when there is degeneracy of the zero-field states. Such examples are legion in all branches of physics. In a classical example, planetary orbits that would otherwise close

Figure 5.14 A statue of Hinduism's Ardhanarishvara, a representation/unity of the male god Shiva with his female principle Parvati, as an illustration of broken symmetry. A chola bronze statue from the 11th century AD. Saline Hansen <http://en.wikipedia.org/wiki/Ardhanarishvara>.

if the Sun were the only other object in the Universe (Sec. 1.2.5) do not do so because of the presence of other planets or corrections from Einstein's General Theory of Relativity. Even for Jupiter, the largest planet, though still small in mass compared with the Sun, all these symmetry breakings are small in this classical example.

Spontaneous symmetry breaking is also easy to comprehend. Even though the equations of motion or Lagrangians and Hamiltonians may possess some symmetries, solutions of those equations describing some system may not display those symmetries. A familiar example is any object that is not spherically symmetric, a pencil say, that points in a preferred direction. This is seen to be just a consequence of choosing one among an infinity of solutions by holding the pencil in one particular direction. The Hamiltonian of all the individual atoms and molecules in the pencil may be spherically symmetric but the global solution does not have to exhibit it. Note that in all such cases there is an infinite degeneracy among the possibilities, the pencil held in any direction having the same energy and other properties. There is no energy cost in replacing one of these degenerate states by another. Grander examples occur throughout physics, and are referred to as zero-energy or zero-mass excitations, called Nambu[10]–Goldstone[11] modes.

Broken symmetry, when slight and with no obvious weak external agent to account for it, poses more significant questions. The departure of the g-factor of an electron from its Dirac value of 2 is small and of the order of the fine-structure constant, which is a measure of the strength of electromagnetism. Thus, QED corrections would be expected to be of this order of smallness. Similarly, in the triplet of elementary particles called pions, the two charged ones (whose masses have to be identical from CPT invariance) have a mass that is of this same order of smallness different from the mass of the neutral pion, again as could be expected from the fact that this difference stems from electromagnetic interactions in which they differ, all else being common to the triplet.

One of the more curious instances of an intrinsic and fundamental symmetry that is slightly broken is that of time reversal invariance. Only a few weak interactions, and even they only slightly, break this invariance [23], which is otherwise valid throughout microscopic physics (the question of the one-way direction of time in our macroscopic experience is different and will be considered in Chapter 7). Why should this be so of a fundamental symmetry of nature, that it is broken but then only very slightly? At least for now, physics reverts to the poetic metaphor of the famous physics master of symmetries, Richard Feynman, the following from his *Lectures in Physics* [9]:

> Why is nature so nearly symmetrical? No one has any idea why. The only thing we might suggest is something like this. There is a gate in Japan, a gate in Neiko, which is sometimes called by the Japanese the most beautiful gate in all Japan; it was built in a time when there was great influence from Chinese art. The gate is very elaborate, with lots of gables and beautiful carving and lots of columns and dragon heads and princes carved into the pillars, and so on. But when one looks closely he sees that

---

[10] Yoichiro Nambu, 1921, Japanese and American. Theoretical physicist with many contributions to elementary particle physics and to broken symmetry in superconductivity and particle field theories such as chromodynamics.

[11] Jeffrey Goldstone, 1933, British. Theoretical physicist, known for his discovery of zero-mass excitations as a result of spontaneous symmetry breaking.

in the elaborate and complex design along one of the pillars, one of the small design elements is carved upside down; otherwise the thing is completely symmetrical. If one asks why this is, the story is that it was carved upside down so that the gods will not be jealous of the perfection of man. So they purposely put an error in there, so that the gods would not be jealous and get angry with human beings. We might like to turn the idea around and think that the true explanation of the near symmetry of nature is this: that God made the laws only nearly symmetrical so that we should not be jealous of His perfection!

It is also important never to forget that physics is ultimately an experimental subject. Whatever the bias and urge we may have to see symmetry as aesthetically beautiful, while it is a guiding principle in approaching the subject, we have to be prepared to accept that the world or Universe is what it is as finally measured by our observations and experimental apparatus. Thus, for many centuries it was held, even by Galileo himself, that circles must describe the orbits of planets, the less symmetric ellipse being 'uglier'. This finally had to give way to Kepler's actual observation, later vindicated by Newton's theory, that a $1/r$ potential leads to ellipses in general, with a varying ellipticity parameter, the zero ellipticity limit being a circular orbit but having no special status. Here again is an element of physics, as in Galileo's own realization, with zero and any non-zero but constant velocity being on the same footing, that the zero value is but one among all possible values and has no special distinction.

So, too, with our modern quantum instances, whether of slight violation of time reversal invariance or a maximal one of parity invariance. Further, whatever our bias in seeing integers as specially distinguished, the g-value of an electron is not exactly 2, and the inverse of the fine-structure constant, α, is not exactly 137, differing by less than 1%. These are facts of our Universe, to be accepted. In the first case, our theory (quantum electrodynamics) gives an account of the many decimal places to which g has been measured; in the second, α is one of the dimensionless constants characterizing the Universe as it is. Perhaps a later theory will account for it but there will then be some other initial inputs to be regarded as given within the physics of that later day.

It seems merely silly of arguments such as the anthropic principle to attribute any value to the fact that we are here (or, more generally, to life being here), and able to pose questions about the Universe. Even granting the premise that except for a narrow band of values, any other α would mean a Universe with no stars or the stars all burnt out very rapidly, and thus in either case not looking anything like the Universe we live in, the argument fails on its own terms to 'explain' the particular value, down to many decimal places, that our experiments measure. Indeed, in even the smallest interval of what is seen as an allowed or permissible band of values lie an infinity of numbers and one has not 'accounted' for the particular one among them that is measured. In invoking an extraneous (to science) appeal to our being here, the anthropic principle prematurely shuts off the quest within science itself to account for the value of α, even should this lie far in the future.

Dirac maintained that it is more important to have 'beauty in one's equations' than anything else. The idea was spectacularly successful for him in his derivation of the equation for the electron, that compact and elegant equation not only reconciling quantum mechanics with Special Relativity but containing so much more within it, including the electron's intrinsic spin angular momentum and its correct coupling to the orbital magnetic field, and the description of the anti-particle, the positron. Symmetry and such aesthetic readings into it are guiding principles in how we work, but we also have to be wary, even in the light of successes when it has indeed led to progress. Einstein, in formulating his General Theory of Relativity, with just two terms, one on each side of the equation, both tensors of second rank, on the one side from space–time curvature and on the other the energy-momentum tensor, dropped another term that could also be admitted into the equation. That matter and geometry could be related down to the most economical and spare capsuling of them in just two terms had a compelling logic and attraction.

The third term involves the so-called 'cosmological constant', Λ:

$$R_{\mu\nu} - g_{\mu\nu} R/2 - \Lambda g_{\mu\nu} = -(8\pi\, G/c^2) T_{\mu\nu}, \qquad (5.1)$$

where $g_{\mu\nu}$ is the metric tensor (Sec. 1.2.4), $R_{\mu\nu}$ the Ricci[12] tensor, a contracted form of the Riemann[13] tensor, itself formed out of derivatives

[12] Gregorio Ricci, 1853–1925, Italian. Mathematician with contributions to algebra and analysis, and inventor of tensor calculus.

[13] Bernhard Riemann, 1826–66, German. Famous mathematician, known for geometrical investigations of curved surfaces, later to become a key part of Einstein's General Theory of Relativity. He made fundamental contributions to number theory as well; his famous 'Riemann hypothesis' is still unproved. It is seen as central to a host of results, including the distribution of prime numbers.

of the metric tensor, R is the trace of the Ricci tensor (that is, the sum of the diagonal entries), while $T_{\mu\nu}$ is the energy-momentum tensor. All quantities on the left-hand side have to do with space–time geometry, whereas that on the right describes the matter that determines that geometry and is in turn influenced by it.

The Λ cosmological term had a curious gyration in its history. Einstein saw correctly that it could accommodate a repulsive gravity, for which there was no evidence in his time, and a Universe that would be expanding under its influence, again with no extant evidence. He dropped it to obtain a static Universe. Yet, it was not too long afterwards that the expansion of the Universe was indeed discovered and the realization that even with just the two terms alone such solutions to the equation existed. The expansion could thus have been seen as a prediction of the theory, leading to the oft-quoted 'greatest blunder' of his to have thought only in terms of a static Universe. And, very recently, the observational discovery that the expansion is itself accelerating has resurrected the cosmological constant as one natural way of accommodating this acceleration. All other considerations of beauty, elegance, and aesthetics must ultimately give way to unambiguous observation and experiment. That is the nature of physics, initially termed 'experimental philosophy'. Just as with Newton's gravitational constant, G, or the speed of light, c, Λ is simply a given constant of our Universe. It is what it is, and for us to measure and incorporate into our physics.

6

Maps in Various Forms

6.1 Maps in Human History

As with most of the themes discussed in this book, maps and map-making go back to early human history, perhaps even further. Along with the idea of a linear dimension for distances, early hominids may well have used primitive forms of maps to keep track of where they were, and recognized landmarks and orientations with respect to the Sun and Moon for returning from the hunt. Even animals and insects use some form of maps. We know that bees use a representational map on the walls of their hives to communicate to their hive mates the location of a source of nectar or pollen, and homing pigeons, monarch butterflies, and long-distance migratory birds are expert navigators.

Seafarers such as the Chinese, or Indians and Arabs whose dhows plied between their lands, must also have had map knowledge of their shores, and they helped the first European sailors to make the crossing from Africa to Asia (an Arab pilot, Ahmad Ibn-Madjid[1], is said to have taken Vasco da Gama[2] in 1498 on the crossing to India). Captain Cook[3],

---

[1] Ahmad Ibn-Madjid, 1421–1500, Arab. Poet and navigator, said to have operated around Oman and the waters of the Arabian Sea.

[2] Vasco da Gama, 1469–1524, Portuguese. Explorer and navigator, and the first European to complete an ocean voyage to India. This and a second voyage opened the spice trade from Asia to Europe. He was appointed Viceroy of Portuguese territories but on his next trip he died of illness in India.

[3] James Cook, 1728–79, English. Explorer, navigator, and cartographer of the British Navy, who first mapped the St Lawrence River before making voyages to the Pacific Ocean, and the first circumnavigation of New Zealand. The first of the Pacific voyages had the scientific objective of observing the transit of Venus, and carried the botanists Joseph Banks and Daniel Solander, whose illustrations and collections of the unique flora of Australia achieved renown. On the second voyage, he carried the chronometers of watchmaker Harrison that decisively settled the problem of locating the longitude at any location, a problem that had been considered the central scientific question for over a century. During the third expedition, he went on to map the north-west coast

when he sailed with a Polynesian priest, Tupaia[4], who directed him between islands not known to Cook, marvelled that he seemed to have all the maps in memory, as did the Polynesian sailors who had traversed the South Pacific for over 2,000 years. With the advent of European navigators from the 15th century on, maps of the continents they travelled to gained major importance and became prized commodities, treasured and fought over by nations and kings.

Maps, and old maps in particular, still hold our fascination (Figure 6.1). They decorate our walls at homes and businesses. These old maps themselves become a story of the development in time of our knowledge of our world. Today, simple geographical maps have evolved into all kinds of representational maps displaying, in an effective visual

Figure 6.1 1522 world map of Laurent Fries, based on the Waldseemüller map of 1513, one of the earliest maps. <http://commons.wikimedia.org/wiki/File:Fries_worldmap_1522.jpg>.

of North America and Alaska. Later, in an altercation with the King of Hawaii and his men, he was speared to death on a beach.

[4] Tupaia, 1725–70, Polynesian. Navigator and tribal priest, whose astonishing knowledge of hundreds of Pacific islands and the surrounding waters, which he had learnt from his father and grandfather, led to his being taken on by Joseph Banks on Captain Cook's first Pacific voyage. He died of illness on board the ship in Indonesia.

manner sometimes more efficient than words, some aspect of our lands and our lives, whether the distribution of wealth or of disease [29]. Figure 6.2 gives an example of what is now available for each disease. And, with global positioning systems (GPS) and hand-held devices, we can carry a wealth of information in the palms of our hands, and know our precise location even in the deepest jungles of the Amazon or out in some vast expanse of an ocean. All this is a 'world away' from the experiences of a Columbus[5], Drake[6], Magellan[7] or Cook and their

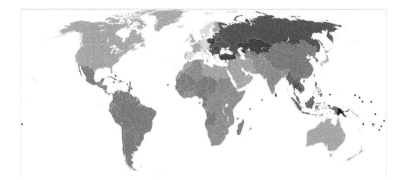

Figure 6.2 A map of musculoskeletal disease, with increased shading (or colour coding from yellow through orange to red) indicating increasing burden as measured by disability-adjusted life years. <http://commons.wikimedia.org/wiki/File:Musculoskeletal_diseases_world_map_-_DALY_-_WHO2004.svg>.

[5] Christopher Columbus, also Colon (Spanish), and Colombo (Italian and Portugese), 1451–1506, Italian. Navigator and explorer, often described as the European discoverer of the Americas through his four expeditions across the Atlantic Ocean under the banner of the Spanish kings. While others, including Europeans, had preceded him, his voyages had a profound impact through Spanish colonization, and through the commerce and interchanges between Europe and the Americas. While his original intent was to discover a route to the spices and silks of Asia by sailing west from Europe, he underestimated the size of the Earth and was unaware of the New World in between, but his voyages had dramatic consequences in the travel in both directions of people, animals, diseases, and food products (corn/maize, cocoa, vanilla, potato, and tomato all originated in South America).

[6] Francis Drake, 1540–96, English. Sea captain, navigator, pirate, and explorer, the second person to head a voyage to circumnavigate the globe. He played a decisive role in the battles between the English and Spanish armadas.

[7] Ferdinand Magellan, also Fernando (Spanish) and Fernao (Portuguese),1480–1521, Portuguese. Navigator and explorer, who got the Spanish Crown to support his

crews, and of countless and unnamed sailors before, who discovered so much of this in the past five centuries. Many, of course, died, and shipwrecks still litter distant shores.

A map, in its simplest and original form, represents the positional layout of objects in a geographic landscape. It has, however, expanded to become a major metaphor of our languages, as a representation of much else [30]. This includes the use of representations in the sense of physics, as will be considered in this chapter. In that usage, there are many dimensions to the word 'representation'. A first parameter about any geographical map is the scale, for instance 1:10,000. This means every unit distance on the map is a faithful rendering (representation) of 10,000 units on the surface of the Earth. Clearly, the smaller that number, the more detail that the map is capable of describing (Figure 6.3). Of course, it would be unrealistic to bring the scale down to 1:1, the map then being essentially the full surface itself or whatever piece of it is being described. Thus maps may approach the underlying reality being described but are not identical to it. So too in physics, which, as a subject, is itself a model for an underlying reality, and we should always be aware that we may hope to get closer and closer but not mistake our descriptions for that reality itself.

The next parameter of a map, again because it is a representation and not the underlying geography itself, is that a map renders on a two-dimensional flat piece of paper or parchment a curved two-dimensional surface of our globe that is embedded in three dimensions. This necessarily introduces certain incompatibilities or distortions, and early history threw up alternative representations, each emphasizing or being accurate for one purpose or another. Among these 'projections', some preserve direction, useful in navigation for setting the compass for travel to a specific destination, but at the price of distorting areas, or vice versa. Indeed, these are incompatible choices, so that it is necessarily so that compass direction and area-preserving maps will violate each other. Of course, a sphere itself, the globe of our childhood possessions, in being geometrically similar to the Earth (although even there our

discovery of the passage between the Atlantic and Pacific Oceans, those Straits now named for him. He observed the Magellanic Clouds, which are dwarf galaxies. While one ship of his fleet completed the expedition, becoming the first to circumnavigate the Earth, and returning with spices from the East Indies, he himself was killed on an island of the Philippines.

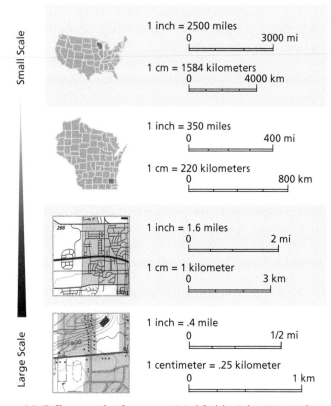

Figure 6.3 Different scales for a map. Modified by John Krygier from John Krygier & Denis Wood, Making Maps 2nd ed., Guilford Publications, 2011. <http://krygier.owu.edu/krygier_html/geog_222/geog_222_lo/geog_222_lo04.html>.

Earth is not a perfect sphere but has the shape of what can only be described in its own terms as a 'geoid'), does not have these distortions but then is often unwieldy for use as a map. Among the major projections that most are familiar with is the 'Mercator'[8], named for its inventor

---

[8] Gerardus Mercator, 1512–94, Belgian. Mathematician and cartographer. He was the first to use the term 'atlas' for a collection of maps. He produced maps of Europe and the world, becoming the leading mapmaker of his age, his son continuing his work after him.

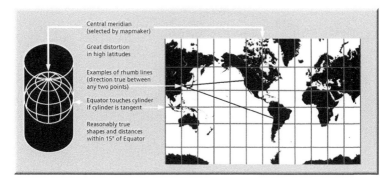

Figure 6.4 The Mercator projection for a map. While preserving compass direction, and thus being invaluable for navigation, areas are distorted, especially at large latitudes. <http://upload.wikimedia.org/wikipedia/commons/thumb/6/62/Usgs_map_mercator.svg/2000px-Usgs_map_mercator.svg.png>.

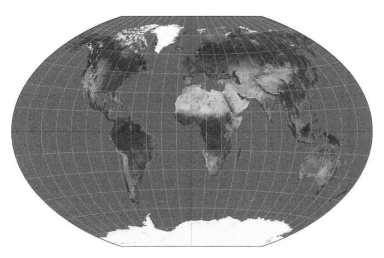

Figure 6.5 A world map. Strebe: http://en.wikipedia.org/wiki/File:Winkel_triple_projection_SW.jpg.

(Figure 6.4). A modern map of the world is shown in Figure 6.5, and the ultimate in maps, of our entire Universe, is now available (Figure 6.6), including three-dimensional renderings of it <http://blogs.discovermagazine.com/outthere/2013/06/16/the-most- amazing-map-youll-see-today-no-matter-what-day-it-is/#.UdOhmhYldzZ>.

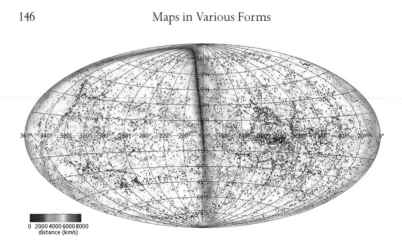

0 2000 4000 6000 8000
distance (km/s)

Figure 6.6 Map of the Universe, displaying all observed galaxies, from H. M. Courtois *et al*, *Astronomical J.* **146**, 69 (2013).

6.2 Maps in Mathematics

We begin with the use of maps in mathematics. A map is an association of one number or entity with another. A mathematical function, $f(x)$, familiar from school algebra or calculus, is thus a map, yielding for each value of x a value (or more than one, for multiply valued functions) taken by that function at that point; thus, $M : x \rightarrow f(x)$. x and $f(x)$ need not be restricted to real numbers but may be more complicated objects themselves. Indeed, a familiar example today on TV or movie screen is a globe 'morphing' into a flat opened map, or equivalents with other physical or biological objects; this is an instance of a map, albeit more complicated.

As an example, the operation of replacing an equilateral triangle, as in Figure 6.7, by the next object in the figure with a triangular projection built on the mid-third of each side, is also a map. Iterating such a map gives the next object in the figure after two iterations, or the last one in Figure 6.7 after several, and generates what is called a 'Koch[9] snowflake' in the limit of infinite iterations. This is an example of what is called a 'fractal', its mathematical dimension lying somewhere between the 1 of the perimeter in any of the finite iterations, and the 2 of the area enclosed. Other fractals, which have fired both mathematical

---

[9] Niels Fabian Helge von Koch, 1870–1924, Swedish. Mathematician known for contributions to number theory and the Riemann hypothesis, the most famous open problem in mathematics (see footnote 13 in Chapter 5).

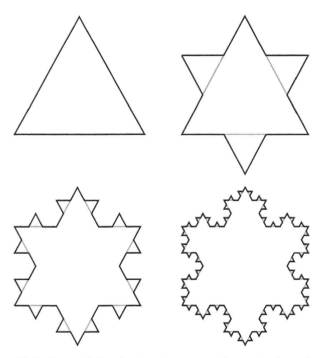

Figure 6.7 Koch snowflake, showing three steps of iteration. In each, a line segment is split into three, with the middle piece projecting out triangularly. In the limit of infinite iterations, a fractal object is generated. David Price <http://upload.wikimedia.org/wikipedia/commons/d/d9/KochFlake.svg>.

and popular imagination, can also be approached most simply through such iterated maps (Figure 6.8).

Although fractals were understood by its mathematical pioneers, Fatou[10] and Julia[11], already a century ago without any such explicit geometrical renderings, it is the advent of modern computers, which make such iterations extremely easy, that has led to the spectacular images that are now so familiar around us. Benoit Mandelbrot[12], in particular, has made 'fractal' an everyday word. Nature, of course, again

[10] Pierre Joseph Louis Fatou, 1878–1929, French. Mathematician and astronomer, known for his work in celestial mechanics and on analysis, especially on bounded analytic functions. One of the first to study what today are called fractals.

[11] Gaston Maurice Julia, 1893–1978, French. Mathematician whose work on the iteration of rational functions is recognized today as early work on fractals.

[12] Benoit Mandelbrot, 1924–2010, French and American. Mathematician who developed the concept of fractional dimensions and studied invariants under

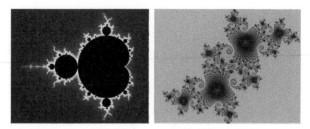

Figure 6.8 Examples of fractals: Mandelbrot and Julia sets. <http://commons.wikimedia.org/wiki/File:Julia_set_(indigo).png?uselang=en-gb, http://commons.wikimedia.org/wiki/File:Mandelbrot_set_1250px.png?uselang=en-gb>.

Figure 6.9 Fractal-like objects in nature: a cauliflower and a fern, along with a fractally generated cauliflower. AVM <http://commons.wikimedia.org/wiki/File:Cauliflower_Fractal_AVM.JPG>.

because iterations are natural for cell division or growth, has near-fractal shapes (Figure 6.9), even though not strictly so in mathematical terms, as physics (ultimately atoms represent a finite end to unlimited division of matter) and biology do not permit that infinite limit to be reached.

transformations in the complex plane, coining the name fractals and popularizing it through incredibly intricate self-similar patterns generated through computer iterations.

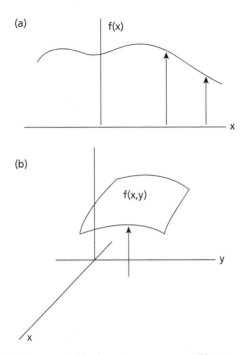

Figure 6.10 (a) A one-variable function as a map. (b) A two-dimensional curved surface as a map.

But we will confine our discussion to functions as simply understood. The one-variable map we started with, $M : x \rightarrow f(x)$, may be represented as in Figure 6.10a. It maps points along the straight line of the horizontal axis onto an arbitrarily curved line that is the collection of the values $f(x)$. A (real) function of two (real) variables, $f(x, y)$, such as the many examples considered in earlier chapters, whether $f(x, y) = x^2 + y^2$ or another, similarly maps any point (x, y) in the horizontal plane onto a number on the surface 'hovering' above it that represents the function $z = f(x, y)$ as illustrated in Figure 6.10b.

Another possible map, or projection, in Figure 6.11 associates each point on a circle with a point on the real line obtained when connecting it to the North Pole, N, and extending it backwards to intersect the x-axis. The North Pole itself becomes the point at infinity in such a projection (note positive and negative infinity become indistinguishable) and is sometimes dropped from consideration, for technical reasons.

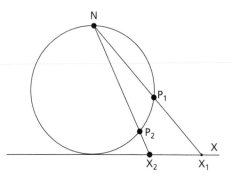

Figure 6.11 Projection of a circle from the North Pole, N, to the horizontal axis, associating an x_i value with each point, P_i, on the circle.

Figure 6.12, which extends Figure 6.11 into an additional dimension, with the circle replaced by a sphere (say the Earth) and the real line by the plane through the Equator, is familiar as a 'stereographic projection' and plays an important role in map-making and in physics. The two-sphere called S^2 is mapped onto the plane called R^2 (a product of two real lines). The problem posed by the North Pole or points near it in the previous paragraph becomes even more acute, and it is clear immediately that areas in that polar vicinity are grossly distorted and spread over large areas in the plane. This is seen in standard geographic maps that have northern portions of Greenland or Siberia spread out so as to look much larger in proportion to other areas on the Earth's surface.

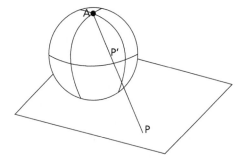

Figure 6.12 Stereographic projection of a sphere onto a plane. This is a higher-dimensional version of Figure 6.11.

Although Greenland is only as large as Brazil, it looks much bigger on such a map (Figure 6.4).

6.3 Maps in Physics

As already seen in other chapters, since transformations play a role throughout physics and a map is a transformation, much of physics can also be viewed as mapping one state of the system to another. This is true of classical dynamics, which, through Newton's laws or Lagrangian equivalent, predicts the state at a later time from knowledge of the state at an initial instant. All classical mechanics is, therefore, a map, initial states mapped onto final states. So too in quantum physics, with again an essential element of all physics being to predict a subsequent state of a system from its state at $t = 0$, this time through quantum equations, which are as deterministic in this regard as are their classical counterparts. The Schrödinger or Dirac equations determine $\psi(t)$ unambiguously from $\psi(0)$ given the potentials governing the motion. The unitary evolution operator, a solution of the equation of motion, performs this map.

It was noted in Chapter 2 that quantum physics brought even more to the fore alternative representations than was already true in classical physics because of the additional feature, from the start, of incompatible choices among conjugate entities, such as the coordinate or momentum representations. This is accommodated naturally in the language of maps, where, for instance, an incompatibility between the preservation of area and direction is already inherent in projecting a two-sphere onto a plane. This goes even further because the very non-locality of quantum physics fits into a similar non-locality in any map of the Earth, this time because a finite curved surface of a sphere is being projected onto an open-ended plane, requiring left and right edges to the map that are artificial. These edges can be placed anywhere and as convenient, but placed they must be and the two edges then identified in the mind so as to wrap around in a continuous traversal from east to west (or vice versa), as done by a circumnavigator.

In a map as in Figure 6.4 or Figure 6.5, two points on the Earth's surface on either side of that edge, are seemingly far apart and not 'local'; this so-called 'adjacency' problem is familiar in cartography [30]. There is an interesting metaphor here for the inherent non-locality of quantum physics because the state of any system with more than one particle resides in a large-dimensional space, not the three-dimensional one of our experience. Even a simple two-electron system such as the helium atom that was discussed in Chapters 2 and 3 has its spatial wave function, $\psi(\vec{r}_1, \vec{r}_2)$, of the two electrons in six-dimensional space, so that we have non-locality in terms of the one-electron spatial probability density and, therefore, any one-electron properties that we measure in our three-dimensional world. With more particles, the situation gets only more complicated, but the non-locality of ordinary mapping of S^2 embedded in a higher (third) dimension onto a two-dimensional plane is an apt metaphor for these larger non-localities of quantum physics.

Among other examples of non-locality, a very illustrative one is called the Bohm[13]–Aharanov[14] effect. An electron passing by a region of magnetic field, $\vec{B}$, has its wave function pick up a phase proportional to the integrated vector potential, $\int \vec{A} \cdot \vec{dl}$, seen along its path. The path may not pass through the magnetic field itself but it is the vector potential, with $\vec{B} = \nabla \times \vec{A}$, that enters in the expression for the phase, and it is $\vec{A}$ that may be present along the path. Given the gauge degree of freedom with different $\vec{A}$ having the same curl describing the same magnetic field, this poses the question of whether to believe in its 'reality' or that the wave function can still sense non-locally the 'real' magnetic field behind that $\vec{A}$.

The question becomes one of experimental physics in an arrangement such as that shown in Figure 6.13, when the electron between source and detector has two alternative paths around a region of confined $\vec{B}$ (see also Figure 8.2). While paths have no meaning in quantum physics, the relative phase difference between the two alternatives does. It is given by the integrated $\vec{A} \cdot \vec{dl}$ in the area enclosed by the loop. This has, however, an unambiguous meaning, being, through Stokes's theorem (Sec. 1.2.2), the surface integral of $\vec{B}$ over that area. That non-zero

[13] David Bohm, 1917–92, American and British. Theoretical physicist who contributed to quantum philosophy and neuropsychology. His alternative interpretation of quantum mechanics through so-called hidden variables and reformulation of Einstein–Podolsky–Rosen's critique of the standard interpretation inspired the work of John Bell. Because of his political views and affiliations, he had to leave the United States in the 1950s for Brazil and England.

[14] Yakir Aharanov, 1932, Israeli. Theoretical physicist, known for his work on topological aspects of quantum mechanics and field theories, and quantum measurements.

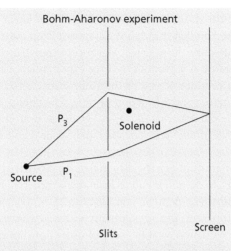

Figure 6.13 The Bohm–Aharanov set-up, showing the two-slit diffraction pattern of an electron beam in the presence of a solenoid holding a confined magnetic field between the slits. <http://www.encyclopediaofmath.org/index.php/Bohm-Aharonov_effect>.

values of such a magnetic field and its flux enclosed within the loop cause interference effects for the electron pattern detected has been experimentally demonstrated. That the electron nowhere encountered the shielded field poses the question of interpretation, that indeed non-local effects of $\vec{B}$ are inherent to quantum physics.

This is also immediate in Feynman's path-integral formulation, where, between source and detector, all possible paths, not just the two or few we think of as natural from classical intuition, are involved and, in this case, these include paths through so-called shielded regions as well. When there is a non-zero magnetic field there, it does have observable effects. Counterparts of the Bohm–Aharanov effect involving magnetic moments moving around a line distribution of charge have also been experimentally demonstrated, all of which reinforce the inherent non-locality of quantum physics.

The Bloch sphere noted in Sec. 4.2.2 for a quantum coin is an S^2 two-sphere and provides an interesting example of one more aspect noted in this chapter, namely, the stereographic projection, or, rather, its inverse, the inverse stereographic projection from the plane, regarded now as the complex plane, onto a two-sphere. The unitary evolution of the state of the quantum spin according to the linear Schrödinger equation can be handled through the standard Pauli matrices, three in number, that describe the $su(2)$ Lie[15] algebra of a spin-1/2. One writes the evolution operator as a product of three exponentials, each having in its exponent a product of one of the Pauli matrices $(\sigma_+, \sigma_-, \sigma_z)$ with the unit imaginary i and a function of time. The choice of this triplet rather than the Cartesian set $(\sigma_x, \sigma_y, \sigma_z)$ gives less non-linearity (quadratic rather than infinite series in sines and cosines) and a more ready interpretation.

The function multiplying the diagonal σ_z is a phase that can take any value from zero to ∞ and is said to have the symmetry algebra $u(1)$ or of the Lie group $U(1)$. A single complex function, $z(t)$, and its adjoint appear as the functions in the exponents with $\sigma_\pm$. This $z(t)$ obeys a so-called 'Riccati[16] equation' that has second-order non-linearity, which is not surprising because writing the wave function as an exponential is a non-linear operation. This first-order differential, but second order in non-linearity, equation is equivalent to the second-order differential but linear Schrödinger equation. Invoking now the inverse stereographic projection of Figure 6.12, the function $z(t)$ can be mapped onto the S^2 Bloch sphere. Together, this S^2 sphere, called the base manifold, and the $U(1)$ phase at every point on it, referred to as the fibre, represent the $SU(2)$ group of spin-1/2 as a 'fibre bundle'. This is a simple way of understanding the role of the Bloch sphere in casting the quantum evolution of a spin into that of a classical unit vector's rotation on a sphere.

The inverse stereographic mapping for spin-1/2 or a two-level system or qubit finds more general mappings when extended to higher spins or N-level systems or qudits of dimension $d = N$. In a very similar construction as in the previous paragraph for $N = 2$, with a product of three operators, the last diagonal, and the others involving sets of complex numbers, $z_i(t)$ can be mapped onto higher-dimensional and

---

[15] Marius Sophus Lie, 1842–99, Norwegian. Mathematician whose extraordinary studies of continuous symmetries and their role in geometry and differential equations permeate those subjects and theoretical physics. An associate of Felix Klein, they together established most of the subject of group transformations. 'Lie groups' and 'Lie algebras' are used throughout quantum physics.

[16] Francesco Riccati, 1676–1754, Italian. Mathematician and engineer who designed canals and dikes in Venice, but known today especially for his studies of differential equations with quadratic non-linearity.

more complicated base manifolds and fibres described by larger groups than $U(1)$. Thus, applied to a pair of qubits and their group $SU(4)$ (see Sec. 4.2.3), analogous but more complicated manifolds to the single qubit Bloch sphere in Figure 4.7 can be described [22]. Also, some of the sub-groups of $SU(4)$, and the corresponding physical systems and Hamiltonians of two qubits can be mapped onto projective geometric designs. Notably, sub-groups involving 7 and 10 out of the full set of 15 operators that occur in many logic gates and 4-level systems can be identified with the diagrams in Figures 5.7 and 5.6, respectively [22].

7

The Problem of Time

7.1 Time in Our Lives

As with the concept of space and our location in it, time has always had
a central hold on humans and human history. The inexorable passage
of time from the moment of our births so dominates our thinking and
language that it is difficult to construct a couple of sentences without
words that are related to it (moment, days, before, after, . . .). But, from
the beginning, the concept of time has posed questions and problems to
lay people, philosophers, and physicists alike. What is time apart from
what our watches and clocks display? Is it universal, a characteristic of
our Universe, or is our experience of time individual to each one of us?

Einstein's Special Theory of Relativity, in fundamentally changing
Newtonian ideas of space and time, which had time as a background
against which we observe motion, showed how even concepts such as
simultaneity of events depend on the observer's frame. Different inertial
frames in relative uniform velocity with respect to one another will differ on such notions (Sec. 1.2.4), making scientific a theme that jokes and
myths have long played with. Thus, the one about a turtle that walks
into a bar and calls for a stiff drink because it has just been mugged
by two snails. To the bartender's solicitous 'What happened?' it says, 'I
don't know, it all happened so fast'. Or a story from Indian mythology
of a king and companion coming to a lake, the king going in for a dip
to emerge minutes later as the companion sees it but the king himself
going through several life cycles and covering the full gamut of human
experience as he sees and experiences it in that 'same' period.

And then there is the old saw about time being a way of keeping everything from happening all at once. Several ancient cultures
regarded time as an illusion. All these find assonances in physics.
For a light beam, as follows from the Special Theory of Relativity,
time stands still. Classical physics already, and even more quantum
physics, admits alternative formulations, one time dependent and the

other independent, raising the question of whether time is a necessary concept or, perhaps, should be done away with all together. This idea has occurred to many and, most notably, has been advocated in physics by Julian Barbour[1].

There are two aspects to time that are conceptually and operationally very different. One is the Newtonian concept of time as a background, uniform flow in one direction from past into present and on to the future. The other is of periodic phenomena where something repeats. This latter, especially when it is regular, provides the way of measuring time intervals. Days and years follow from the motion of our Earth: rotation about its own axis, and revolution in orbit around the Sun, respectively. The human heartbeat is another, this time on our own scale, and approximately the second that results when a day is divided by definition into hours, minutes, and seconds, the conversion numbers of 24 and 60 being historically set.

As in the example presented in Sec. 1.1, a pendulum of 1 m oscillates from one end of its swing to the other in something very close to a second because of the particular value of Earth's gravity, g (note that this depends mainly on the mass and radius of the Earth, not its motion, although the effective gravity at any location is slightly affected by the rotation and its attendant centrifugal acceleration, thus depending on latitude, but the deviation is never more than 0.3%). Indeed, Galileo is supposed to have arrived at this realization by timing against his own heartbeat a censer that he observed swinging in church. With the effects of tides and other variations making the astronomical clock of the Earth's motion not accurate enough for modern science and technology, the second is now defined in terms of something much more accurately periodic, the transitions in an atom as it passes from one energy state into another. Today's time standard is given by such 'atomic clocks' maintained at the National Bureaus of Standards of various countries.

7.2 Time in Classical Physics

Time as an absolute background, the same for all observers, was part of Newton's formulation of classical mechanics. With absolute velocity

---

[1] Julian Barbour, 1937, British. Physicist, interested in quantum gravity and the nature of time.

meaningless, only relative velocities having significance, frames of reference would differ in their ascribing of position, depending on the relative velocity between the frames, but all would share the same t. This is Galilean or Newtonian relativity. A primed and unprimed coordinate system of two such frames with relative velocity $\vec{v}$ have their spatial coordinates related by

$$\vec{r}' = \vec{r} + \vec{v}t \tag{7.1}$$

with the same time for both frames, $t' = t$. The laws of mechanics seemed consistent, especially since they involved only accelerations, and both frames would agree on them, and on the forces and torques present. It was, however, with the advent of electromagnetism, Maxwell's equations not being consistent with Galilean/Newtonian relativity, that a major re-examination was called for, as realized by Lorentz, Poincare[2], and others, and finally and most decisively by Einstein.

Einstein's Special Theory of Relativity modifies these equations to a slightly more complicated mixing up of space and time,

$$\vec{r}' = \gamma(\vec{r} + \vec{v}t), \tag{7.2}$$
$$t' = \gamma(t + vt/c^2),$$

with $\gamma = 1/\sqrt{1-(v^2/c^2)}$, called the Lorentzfactor and responsible for the length contractions and time dilations observed in one frame as viewed from the other. The speed of light, c, being large on the scale of most phenomena encountered at the macroscopic level (an especially fast meteorite that crashed in California in April 2012 was travelling at 30 km/s, still small on the scale of light velocity's 10,000 times larger value, v^2/c^2 effects therefore being only one part in 10^8), relativistic effects did not become evident until the exploration of the microscopic level of atoms. But, with today's accuracy of GPS (global positioning

[2] Jules Henri Poincare, 1854–1912, French. Mathematician, physicist, and philosopher, with many contributions to celestial mechanics (especially the three-body problem), applied mathematics, and mathematical physics. He was one of the founders of topology and chaos theory in classical dynamics, and introduced group theory into physics. He also independently discovered what are called Lorentz transformations and some of the ideas of Special Relativity. A popularizer of mathematics and science, he wrote several books for the layman. He differed from Kantian philosophy, and from Bertrand Russell and Gottlob Frege, in arguing for the supremacy of intuition over logic in mathematics.

system) and other technologies, they, along with general relativistic effects of similar order, are crucial for something as mundane as the location maps on the dashboards of our cars.

The Einstein–Lorentz transformations of Eq. (7.2) give the correct description of all classical mechanics and electromagnetism, including translations, rotations, and velocity boosts between inertial frames. The invariant that follows (Sec. 1.2.4) from these equations is the space–time interval $c^2 t^2 - r^2 = c^2 t'^2 - r'^2$. Classical mechanics and classical electrodynamics are perfectly internally consistent and coherent theories.

Lagrange and Hamilton reformulated Newton's equations in a more general and convenient form. Energies, potential or kinetic, scalar quantities and not vectors such as forces and torques, became the basic ingredient, and the equations were form invariant, the same in any coordinate system (Sec. 1.2.3). Both were great simplifications for the handling of the equations and for finding solutions. Their immediate connection to conservation laws when a symmetry means the absence of a coordinate in the Lagrangian has been discussed in Sec. 5.1.2. But, even more significantly, for the theme in this chapter, they allowed for a 'global' rather than 'local' picture of motion.

Instead of a specification at every instant between the motion's initial A and final B of the particles' positions and velocities, any next instant's values following from the forces prevailing at the previous instant, this global formulation is very different. It considers the entire motion in terms of an 'action', defined by an integration over t of the Lagrangian between A and B, thus involving all times in between. An associated variational principle, that the actual motion is the path that makes this action integral stationary, is a fundamental change of ground philosophically. The local description from one time step to the next is replaced by a global statement about the full motion. That the local and global formulations lead to the same physics is because of the statement of stationarity under *all* possible variations. These would include a path only slightly deviating around some local position from the actual one, so that, without holding locally as well, the global stationarity would fail (Figure 7.1). But it also raises the question of doing away with consideration of time because t is integrated over, and so those intermediate values between t_A and t_B are irrelevant to the physics of the motion.

Next, there is the question of initial conditions. Newton's emphasis on initial conditions is almost as important a contribution of his to

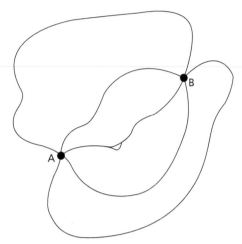

Figure 7.1 Lagrangian paths connecting events A and B, with the classical path shown in the middle, along with a close one that differs only in one tiny segment.

physics as are his laws of motion. He held that these laws were not enough but also necessary are the initial conditions for any physical system. Thus, to those who questioned his theory accounting for elliptical orbits in the Solar System but not the specific ones seen or why they are all in a plane, he said that there are always initial conditions, and a future theory that goes further and explains some of them will have its own conditions that lie outside the purview of the laws themselves. Indeed, Laplace's nebular hypothesis and the flattening of a gas cloud into a disc does account for orbits ending up in a plane, but the initial swirls in the nebula accounting for the angular momentum in the system then become the new initial conditions.

As a simple question of mathematics as well, a differential equation by itself is not enough but initial or boundary conditions are needed to specify any particular solution. For a second-order differential equation such as Newton's, these are often taken as the coordinates and velocities specified at the initial instant, and this is what is meant by a physical system in Newtonian mechanics. The Lagrangian formulation through the stationarity of the action instead holds A and B fixed, the coordinates and times at those boundary points being fixed. We will consider in Sec. 7.4 another formulation through first-order but non-linear equations with again only an initial condition.

Newton, Lagrange, and Hamilton treat time differently from space, which carried over into quantum physics as well, initially in non-relativistic quantum mechanics. The problems this created for quantum physics will be discussed in Sec. 7.3 but it is worth noting how Hamilton in particular was almost prescient in anticipating quantum mechanics, including in formulating the action integral, the dimension of action being that of Planck's quantum constant. With the discovery of that fundamental constant, Feynman's path integral formulation follows almost naturally, by placing the action divided by $\hbar$ for all possible paths in a sum over them, $\exp(iS/\hbar)$, to lead to the quantum wave function. The relativistic generalization, where space and time must be treated on an equivalent footing, as achieved finally in quantum field theory (but not mechanics), was also almost anticipated by introducing an integration over space through a Lagrangian (space) density, both $\vec{r}$ and t thus being integrated over, and thereby dethroned from any central role in physics.

7.3 Time in Quantum Physics

Heisenberg and Schrödinger developed non-relativistic quantum mechanics, with again time treated differently from space. The spatial coordinates were treated as operators along with their conjugate momenta, with which they do not commute. The Born–Heisenberg commutator, $[x, p] = i\hbar$, or the equivalent regarding of linear momentum as a gradient in space, $p = (\hbar/i)\partial/\partial x$ or $\vec{p} = (\hbar/i)\vec{\nabla}$, along with $E = H = i\hbar\partial/\partial t$, leads to the Schrödinger equation of motion that replaces Newton's. This conversion of the expression of the Hamiltonian of a system as a sum of kinetic and potential energies, $H = (\vec{p}^2/2m) + V(\vec{r})$, into the first-order differential in t and second-order differential in space Schrödinger equation satisfied the physical requirement of an equation of motion, that it predict the state of a system at t from knowledge of that state at $t = 0$. But it conflicts with the Special Theory of Relativity in treating space and time differently. Indeed, the Heisenberg uncertainty relationship between x and p that also follows from the above commutator does not have an exact counterpart for energy and time. Although there is an uncertainty link between energy and time, its interpretation is quite different, there being no operator associated with time, t remaining just a parameter in quantum mechanics, as it did in Newtonian mechanics. Attempts to make t an operator have always failed, which is another pointer to the theme of this chapter about doing away with it for physics.

It is a curious fact that Schrödinger initially wrote down a relativistically correct equation, which is not a surprise, as he would have been well aware of Special Relativity's role in physics, but he discarded the resulting equation. Starting with (Sec. 1.2.4) the relativistically correct relationship, $E^2 = c^2 \vec{p}^2 + m^2 c^4$, rather than the non-relativistic expression in the previous paragraph, and making the same replacements of E and $\vec{p}$ by differentials in time and space, led him to an equation but it was also second-order differential in time (just as it was in space). This seemed incompatible with a proper equation of motion which, in predicting the state at a later t from knowledge at $t = 0$, should be of first order. He then formulated the equation that goes by his name, his discarded equation being later recognized as also a correct one, indeed relativistically correct, but describing a relativistic scalar field, not the mechanics of a particle. It is today called the Klein–Gordon[3] equation.

7.3.1 Time in Non-Relativistic Quantum Mechanics

As is its counterpart in classical mechanics, the non-relativistic quantum mechanics of a particle is also internally consistent. Given the wave function, $\psi(0)$, at time zero, it correctly describes the evolution to the state's wave function at a later time, $\psi(t)$, unambiguously, just as Newton's equations describe the evolution of the classical state from the values of its coordinates and velocities at an initial instant. Only what is meant by the state of a system differs between classical and quantum physics. As in the Lagrangian formulation, there is also a variational formulation of the Schrödinger equation.

There are, however, in non-relativistic quantum mechanics, even more sharply than in classical mechanics, two different formulations describing the same (and all) physics, one time-dependent and another time-independent, this latter in terms of the stationary states of the system. This can be illustrated already by one of the simplest applications, namely encounters of a one-dimensional particle with

---

[3] Walter Gordon, 1893–1939, German. Theoretical physicist who worked on the relativistic treatment of quantum particles.

potential steps and barriers, which is taught in the very first lectures to an undergraduate student and also historically important as one of the first successes of quantum physics as applied by Gamow[4] and others to what is known as 'quantum tunnelling'. This is one of the most ubiquitous and important of quantum phenomena. Nuclear fusion in stars, covalent binding of two atoms in a molecule, and the various tunnelling microscopes of today are but some examples besides radioactivity in which tunnelling through a barrier is involved.

Most textbook and other discussions of a particle on a one-dimensional step or barrier (Figure 7.2) are phrased in terms of time, that the particle is incident, say, from the left with some energy $E = p^2/2m$ and is thus described by a travelling wave, $\exp(ipx/\hbar)$. Along with the (often implicit, and not always shown) time dependence, $\exp(-iEt/\hbar)$, this

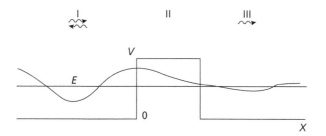

Figure 7.2 A step potential barrier that reflects and transmits waves and particles incident on it from the left. That incident and reflected waves exist in region I and only transmitted waves in region III, also for energy E below barrier height V (quantum tunnelling), is illustrated, as is also the wave function over all x. The wavelength is the same in regions I and III, but the amplitude is lower in the latter, indicating a transmission probability smaller than unity (but not zero) for $E < V$.

---

[4] George Gamow, 1904–68, Russian and American. Theoretical physicist and cosmologist. He was one of the first to recognize the effect known as tunnelling in quantum mechanics, using it to account for alpha-decay radioactivity. He is known for advocating the Big Bang Theory of the origin of the Universe, worked on nucleosynthesis in the early Universe, and predicted the relic cosmological radiation that was later observationally discovered as pervading the Universe in all directions and confirming the Big Bang origins. After the discovery of the structure of DNA, he played a part in advancing the ideas that led to the discovery of the DNA code for amino acids. Known for his playful pranks and puns, he was a popularizer of science with a series of books in which a character called Mr Tompkins encounters relativistic and quantum phenomena in terms accessible to the layman.

represents a particle travelling from left to right, whereas $\exp(-ipx/\hbar)$ would represent the opposite of a particle moving from right to left. Along with the reflected wave in region I in Figure 7.2, both independent solutions in region II, and only a wave travelling from left to right in region III because no particle came at the barrier from the right, continuity at the borders between the regions to get a single continuous solution from $-\infty$ to ∞ gives the desired solution, as shown, from which are extracted reflection and transmission coefficients of the barrier at any energy E. This is a time-dependent description with states $|E, p\rangle$ of energy and linear momentum, and using complex wave functions from the start as solutions of the Schrödinger equation.

As already observed, however, alternative representations are on an equal footing in quantum physics. For a free particle, instead of the eigenstates of the previous paragraph, another such is where the pair of commuting operators is chosen to be the Hamiltonian and parity, as was discussed in Chapter 5. These are standing waves in place of travelling waves, with real sines and cosines instead of the complex exponentials. Any solution for a one-dimensional free particle can be written in terms of them and is equally acceptable. Thus, superpositions of sines and cosines can be written for each of the regions I and III, and either real exponentials (rising and falling) or hyperbolic functions for region II, with continuity again established at the boundaries. The entire discussion can proceed with real functions without introducing any complex elements, which is consistent also with dealing with a real stationary-state Schrödinger equation with real potentials, nothing complex needing to be invoked.

This is also at the same time (!) a time-independent description. It is only the boundary condition at $\pm\infty$, that a particle went in from left, not right, that can introduce complex elements. It is also here that the parity symmetry is broken, at this level of the boundary condition that distinguishes left from right, not in Figure 7.2 or the Schrödinger equation, which are parity invariant. For this reason, the two parity solutions are superposed. But simple analysis [18] shows that the same reflection and transmission coefficients/probabilities follow as in the previous paragraph. Since a superposition in regions I and III of a sine and a cosine can be written as a phase-shifted sine (for odd parity) or cosine (for even parity) due to the presence of the potential barrier, the final expressions for reflection and transmission coefficients can be written as $\sin^2(\delta_+ - \delta_-)$

and $\cos^2(\delta_+ - \delta_-)$, in terms of 'phase shifts' for even and odd parity. It is the difference in the two phase shifts that matters for the observed physics. In the final physics of measurable quantities, the two formalisms are equally valid, equally good.

This extends immediately to the case where the barrier is replaced by a well. Now there arises also the possibility of bound states in that one-dimensional well. With the boundary condition (that wave functions must fall exponentially at $\pm\infty$) also real, it is natural to use the standing waves and real functions of parity eigenstates and get the conditions for the special values of energy where bound states of either parity occur. The condition appears as $\tan \delta = -i$ and thereby as a transcendental equation involving E. Typically, only a few discrete values of E support bound states, whether of even or odd parity. Travelling waves can also be used to set up the condition for bound states, with an imaginary wave vector, $k = i\kappa$, and requiring only falling exponentials at $\pm\infty$, and will give the same result [18].

Thus, as far as physics is concerned, a time-independent representation with standing waves is entirely equivalent to a time-dependent one with talk of waves that travel in time. For the purposes of this chapter, this contrast between two seemingly different treatments of the same physics points to how an analysis with no reference to time works in the quantum world. The key parameters instead are phase shifts that depend on the energy, E, and are measurable quantities experimentally. Not just bound states but even scattering, usually thought of in terms of time and motion, are amenable to a time-independent description. Three-dimensional scattering theory in more complicated situations can also be analysed in the time-independent formalism. Interestingly, time as a conjugate to energy, a key feature of quantum physics with $\hbar$ playing the translating role, can be brought back into the discussion as a 'Wigner time delay' in terms of measurable (experimentally accessible) quantities such as $2\hbar d\delta/dE$.

7.3.2 Time in Relativistic Quantum Mechanics

The previous section noted that the connection between energy and momentum in the relativistic expression did not lead to a differential equation that is first order in time and was initially discarded. To place space and time on an equal footing and have both energy and momentum enter linearly to get a first-order differential in the equation of motion, Dirac factorized $E^2 = c^2\vec{p}^2 + m^2c^4$ by 'taking a square root' of it.

This led him to invoke an internal four-dimensional space so that besides the space–time dependence, the wave function also had these internal four dimensions, represented as a four-column vector. Alongside appeared four-dimensional square matrix operators, the 'Dirac gamma matrices', and the resulting $E = \vec{\gamma} \cdot \vec{p} + \gamma_4 mc^2$, with replacement of energy and momentum operators by first derivatives in time and space, respectively, gave the Dirac equation, $[\gamma_\mu \partial/\partial x_\mu + (mc/\hbar)]\psi = 0$, which is manifestly relativistically covariant. The space–time index, μ, runs over four values and the gamma matrices themselves are 4×4 in an internal four-dimensional space, this coincidence in that number with the four of space-time being accidental.

This was a great triumph, and the transformation properties could be studied of the γ matrices under space–time translations and rotations, Lorentz boosts from one inertial frame to another, and under parity (P), charge conjugation (C), and time reversal (T). The internal four dimensions were interpreted as a combination of two aspects, although it took some time (!) for full appreciation of them. First, one factor of two describes the intrinsic spin angular momentum of 1/2 (in units of $\hbar$), quantum spin thus appearing naturally and not inserted by hand as previously had to be done to accommodate observed atomic spectra.

Further, the g-factor of 2 in coupling spin to a magnetic field, that spin angular momentum couples twice as strongly as an orbitally derived angular momentum, which had been experimentally observed, also came out naturally through the Dirac construction. Spin, therefore, is a relativistic quantum phenomenon. The other two dimensions of the intrinsic space were interpreted as the anti-matter counterpart, whether of electron, muon, or proton, that the Dirac equation also describes and, indeed, for consistency is required to do so. This amounted to a 'prediction' of such anti-particles as positron or anti-proton, the charge conjugates of the previously known particles (the muon comes as a positively and negatively charged pair). The Dirac equation does not describe only an electron but perforce has to include the positron, its anti-particle!

While all this represented a major advance made by Dirac's placing of space and time on an equal footing, it also pointed to an inevitable failing of the quest for a consistent relativistic quantum mechanics of *one* particle. The very fact that such a mechanics had to include the anti-particle and that the marriage of relativity and quantum physics permits the conversion of energy into pairs of particle and anti-particle means that there is no relativistic mechanics of a single particle. The description of *one* electron inevitably includes a cloud of such pairs so that the number of particles is not a conserved quantity or an invariant in a relativistic quantum world. This means that one has to turn to a field theory in seeking such a consistent theory. Besides empirical observations such as the 'Lamb[5] shift' and the small departure of *g* from the value of 2 that also pointed to this, it is clear that while the Dirac equation involves both space and time linearly, it nevertheless still has ('suffers from') the same feature, ever since Newton, of treating time as a parameter and not an operator, whereas space is so treated as an operator conjugate to linear momentum.

7.3.3 Space and Time in Quantum Field Theory

Finally, with the realization that there is no consistent mechanics of a particle when relativity and quantum physics are combined, physicists came to the conclusion that a consistent picture of reality requires field theories. The Dirac equation, or the Klein–Gordon equation, in this view are not equations of a spin-1/2 or spin-0 particle, respectively, but describe corresponding fields, and their solutions are field functions, not wave functions of a particle. Both space and time on which these field functions depend are parameters and not operators. This places space and time on an equal footing, as required by Special Relativity. Both non-relativistic quantum mechanics and Dirac's initial formulation of relativistic quantum mechanics violated this equivalence in regarding only space as an operator, not time. Quantum field theories treat both space and time as parameters, the fields themselves being the operators.

In giving up on making *t* an operator, spatial coordinates too are no longer seen as operators but just background parameters or markers. It is the field functions themselves that are operators, and they are defined at all points of the background grid of space and time. Upon writing a Fourier decomposition of the field functions in terms of plane waves in space and time, the coefficients are operators and are referred to as creation and destruction operators of corresponding field quanta.

---

[5] Willis Eugene Lamb, 1913–2008, American. Physicist whose precision experimental spectroscopy led to the detection of a small energy shift of atomic levels that triggered the development of quantum electrodynamics as the first quantum field theory.

It is they and, therefore, correspondingly, the field functions that obey commutation (for spin integer) or anti-commutation (for spin half-odd integer) rules. A basic feature of quantum field theories is that a consistent quantization procedure requires the use of commutators and anti-commutators, respectively, for fields that describe integer or half-odd integer spins. Correspondingly, the associated particle excitations, which also have the same spins as their fields, are termed 'bosons' and 'fermions', respectively, and obey the corresponding Bose–Einstein and Fermi–Dirac statistics.

Particles themselves are seen as the quanta of excitation of these fields. One need consider only one basic state, the vacuum state with energy, momentum, angular momentum, and all quantum numbers zero, and all physics can be described as expectation values in this vacuum state of operator products of field functions. A field function is an infinite expansion over creation and destruction operators, with multiplicative plane wave factors. A creation operator, denoted as $a^\dagger$, acting on the vacuum ket $|0\rangle$ represents one quantum of excitation. Sandwiching a field function between the vacuum bra, $\langle 0|$, and this ket, $a^\dagger|0\rangle$, results in the quantum-mechanical wave function of the corresponding particle. More complicated products of creation operators acting on the vacuum create multiple excitations. In the case of a nonzero spin, the field function has creation (and destruction) operators of both particle and anti-particle, and pairs of them can also be described in such a formalism. It also becomes natural that even for a single particle, the resulting wave function exists over all space–time, thus accounting for one of the first non-intuitive (that is, on the basis of classical intuition) features of quantum physics of a particle not being localized at a single point in space and time.

With quantum field theory, and the picture of interacting quantum fields, a consistent physics was finally in hand once again after the consistent world view that had been provided by the non-relativistic classical mechanics of particles. The step from the Dirac equation to its interpretation as a relativistic Dirac field and the resulting quantum electrodynamics (QED) also predicted small effects beyond the quantum mechanics of an electron. Thus, in the latter's treatment of the hydrogen atom, the degeneracy between the $2s$ and $2p$ states that goes all the way back to the non-relativistic Bohr–Schrödinger treatment and that persisted through application of the Dirac equation for states with the same total angular momentum, $j = 1/2$, was seen to be lifted. This is the Lamb shift, measurement of it having triggered the formulation of QED by Feynman, Schwinger[6], Tomonaga[7], and others. It is a small but crucial field-theoretic correction, being only 4.5×10^{-6} eV, to be contrasted with a typical non-relativistic energy such as the 10.2 eV Bohr energy difference between this pair of states and the ground state, or even the spin-orbit relativistic correction, the energy difference between $j = 1/2$ and $j = 3/2$, which is 10 times larger.

Since that first successful quantum field theory of QED, many more field theories have been elaborated, such as the electroweak or quantum chromodynamics or the standard model, all having the same structure in terms of their treatment of space and time. For the theme in this chapter, of doing away with time in physics, they introduce now the natural accompanying element that space, too, is but a marker and not a fundamental element of physics. The fundamental elements are energy, and linear and angular momentum, all observables associated with symmetries and laws of conservation. Time and space are to be viewed as derivatives with respect to energy and linear momentum, respectively, together with factors involving the twin elements of i and $\hbar$ of quantum physics (see Sec. 2.2).

7.4 The Invariant Imbedding Approach

Newton formulated the first equations of motion in physics as second-order differential equations with initial conditions on position and velocity. A completely different set of equations follows for a variety of problems, including those of mechanics, in a formulation that views any given problem as part of, 'imbedded in', a family of problems. Thus, consider the question of elementary physics of how high a particle rises when thrown from the surface of the Earth with some initial velocity, v. The conventional Newtonian approach to projectile motion under Earth's gravity, g, is to integrate Newton's equation, $\ddot{x} = -g$, with

---

[6] Julian Schwinger, 1918–94, American. Versatile theoretical physicist and one of the co-developers of quantum electrodynamics. He worked on the development of radar. His masterful use of variational principles and Green's functions originated from such work in electromagnetism but extended to quantum field theories. He was influential through his various books and reports and the many theoretical physicists who were his students.

[7] Sin-Itiro Tomonaga, 1906–79, Japanese. Theoretical physicist and one of the independent co-discoverers of renormalization and quantum electrodynamics.

$x(0) = 0, \dot{x}(0) = v$, to get $x(t) = vt - gt^2/2$ and $\dot{x}(t) = v - gt$, familiar kinematic equations, and determine the maximum height reached as the point where the velocity vanishes, namely, at time $T = v/g$ as given by that second equation. 'Eliminating' T by substituting into the first equation then gives the desired answer, $H = x(T) = v^2/2g$.

The invariant imbedding approach, developed initially by Ambartsumian[8] and Chandrasekhar[9] for astrophysical applications of radiation flow through stellar atmospheric layers, and then as a general technique by Bellman[10], Kalaba[11], and their collaborators, proceeds completely differently. The question of height reached for velocity v, $H(v)$, is seen as a part of a family of such questions for varying v. Among these is certainly the 'trivial' one, $H(0)$, the height reached when thrown with zero velocity whose solution is immediate: $H(0) = 0$. Next, in the particle's rise, when the particle has risen slightly and its velocity dropped to $v - \Delta v$, the height reached from there is part of the family of questions: $H(v - \Delta v)$. The amount risen is easily written as a matter of definition of acceleration, that it is the distance covered at velocity v for the infinitesimal instant $\Delta v/g$, namely $v\Delta v/g$. Thus, clearly (Figure 7.3),

$$H(v) = H(v - \Delta v) + v\Delta v/g. \tag{7.3}$$

The rest is simple calculus to get the first-order differential (note in v, not in time) equation,

$$dH/dv = v/g, \tag{7.4}$$

[8] Viktor Ambartsumian, 1908–96, Armenian. A major figure in theoretical astrophysics with many contributions to the study of stars and galaxies. In studying light diffusion through media for astrophysical applications, he developed the mathematics of invariance principles that later became the general method of invariant imbedding.

[9] Subrahmanyan Chandrasekhar, 1910–95, Indian and American. Astrophysicist and mathematical physicist, with wide-ranging contributions to our understanding of stellar atmospheres and stellar structure. Using quantum mechanics for the electron gas in a collapsed stellar core, he established an upper limit that bears his name for the size of such 'white dwarfs'. Also, he was one of the first to study the quantum mechanics of the negative ion of hydrogen, both for its structure and for its role in stellar opacities. He was the author of many texts on rotating figures of equilibrium, hydrodynamics, black holes and gravitational waves, and a study of Newton's *Principia,* and an influential editor for two decades of the *Astrophysical Journal.* The NASA x-ray telescope is named after him.

[10] Richard Ernest Bellman, 1920–84, American. Applied mathematician, inventor of the method of dynamic programming with practical applications in control theory.

[11] Robert E. Kalaba, 1926–2004, American. Applied mathematician, with contributions to dynamic programming, and communication and control theory.

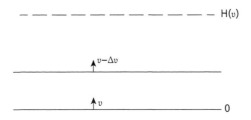

Figure 7.3 The invariant imbedding analysis of $H(v)$, the height reached by an object thrown vertically with speed v against Earth's gravity.

which can be immediately integrated with the initial condition already set for $H(0)$ to give the result $H = v^2/2g$.

This illustration of invariant imbedding is instructive in its comparison with the conventional approach that invokes time explicitly, determines T, and then eliminates it. All this is avoided in the invariant imbedding approach. Even the invoking of the infinitesimal instant, which would seem to introduce the concept of time, is actually for our ease of use of language, $\Delta v/g$ following as a simple consequence of acceleration being a rate of change of velocity, so that the distance covered at speed v while changing by an amount Δv at a rate g is given by $v\Delta v/g$ without reference to time. 'Rate of change' also seems to involve the concept of time, which just goes to show how inescapable it is for us to use words denoting time. But, the important point is that there is no explicit introduction of time that is later eliminated as in the conventional approach.

What is striking about this approach is its economy in focusing on just the quantities of interest, height and velocity, which are at the same time measurable or experimentally accessible quantities at the end points of the motion. While true that with our stop-watches the time, T, is also accessible to measurement, even so it is redundant and not what was being sought in the framing of the problem, namely, the height reached for a given velocity of throw. No time element is necessary for this physics. The very formulation of imbedding in a family of problems and then the use of calculus leads to first-order differential equations and initial value problems rather than the conventional treatment's second-order equations with two boundary conditions.

A quantum-mechanical example is in scattering theory, even a simple one such as scattering by a potential, $V(r)$. The time-independent

Schrödinger approach is to construct wave function solutions of that linear second-order differential equation and fit them asymptotically to a superposition of a standard pair of regular and irregular solutions, typically a sine and a cosine, and thus extract a phase shift relative to the regular solution in the absence of the potential. These phase shifts make contact with experimental data on cross-sections, total or differential in energy or angular distribution, that are expressed in terms of such phase shifts.

The imbedding approach is to think of a family of potentials, $V(r)$, in particular the given potential, as being built up of a series of such, starting at $r = 0$ and truncated at different r all the way to ∞. The phase shift for any one of these is now a function of r, $\delta(r)$. Clearly, with the potential truncated at the origin itself, that is, no potential at all, there is no phase shift, so that $\delta(0) = 0$. On the other hand, given the solution $\sin[kr + \delta(r)]$ at some r, the infinitesimal extra potential (Figure 7.4) between $V(r)$ and $V(r + \delta r)$ will add to it an infinitesimal extra phase shift that can be written simply from the Born approximation

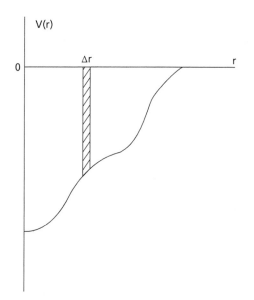

Figure 7.4 The invariant imbedding approach to quantum-mechanical potential scattering from a potential, $V(r)$. The full potential is viewed as being built up of tiny segments, such as the one shown shaded.

($k = p/\hbar$ is the wave vector). Once again, using next the steps of the calculus, a first-order differential (note in r, not in time) equation follows for $\tan \delta(r)$. The Born expression, being a perturbation result, is quadratic in the wave function $\sin[kr + \delta(r)]$ so that the first-order equation is non-linear in $\delta(r)$. Together with the initial condition, $\delta(0) = 0$, the problem is well defined and can be integrated to give the desired $\delta(\infty)$ [31].

The connection between the two approaches is also clear. The wave function at any value of r of the conventional treatment can be written as a phase and an amplitude, both r dependent, and in a so-called phase-amplitude method (PAM), the Schrödinger equation decomposed into a pair of equations for the phase and amplitude functions, as in Sec. 2.1. The first-order, quadratically non-linear equation for $\tan \delta(r)$ that follows is precisely the one given by the imbedding approach. Once $\delta(r)$ is obtained from it, the first-order differential equation for the amplitude function is easily evaluated by quadrature. Again, as in the projectile example, it is inherent to the imbedding method to lead to first-order differential equations, albeit non-linear and often quadratically non-linear. This connects to similar Riccati equations noted in Sec. 6.3 or to the Hamilton–Jacobi[12] equation of classical Lagrangian physics. The commonality is also obvious that in writing a function in terms of a phase in an exponent, such a non-linear operation transforms a linear but second-order differential equation into a first-order but (quadratically) non-linear one for that phase.

Viewing either of the two examples above directly in the imbedding philosophy, this calculus in terms of a family of problems leads naturally to first-order differential (in some relevant physical observable such as v or r of the examples, not in t) equations and initial value problems. There is also an economy in this approach, especially in not invoking elements, whether time or wave functions, that are not accessible or done away with anyway at the end. The definition of acceleration or the perturbative Born result that involves only the square of the wave function (a quantity accessible to measurement) are enough as input to give the full solution. Quantum physics, in particular, brought this accent,

---

[12] Carl Gustav Jacob Jacobi, 1804–51, German. Mathematician with fundamental contributions to dynamics, elliptic functions, and number theory. One of the founders around 1830 of the 'physics seminar' at Königsberg, which combined rigorous training in both theory and experiment, and which shaped curricula in physics first in Germany and then elsewhere.

especially emphasized by Bohr and Heisenberg, of physics dealing only with what is in principle measurable with our experimental apparatus. The imbedding approach fully conforms to this philosophy, developing equations for $H(v)$ or $\delta(r)$ rather than time or wave function. It raises, therefore, the question of physics doing away altogether with 'irrelevancies' such as time (or wave functions).

7.5 Should Time Be Abolished?

Julian Barbour is perhaps the contemporary physicist who has thought most deeply and for long about the nature of time. He, too, favours a time-independent formulation of physics, with time itself regarded in terms of a change in something else: 'Physics must be recast on a new formulation in which change is the measure of time, not time the measure of change'[32]. And, instead of time flow, he would replace by instantaneous snapshots of the Universe as a whole, instants of time that he calls NOWS. Time for him is a stringing together of these NOWS. This suffers, however, from the same profligacy as does the Many Worlds Interpretation advocated for quantum physics (Sec. 8.5). Further, quantum physics makes $\Delta t \rightarrow 0$ problematic because it implies an infinite amount of energy, the conjugate quantity to time. Indeed, even in the two theories of relativity, the same holds true. Any non-zero mass particle can, according to the Special Theory of Relativity, travel at speed c, which is when time stands still, only at infinite energy cost. And, in the General Theory of Relativity as well, where gravitational fields are seen to slow down clocks, the limit of slowing a clock down to zero requires an infinite gravitational potential.

The 17th-century artist Maria Sibylla Merian[13] of Frankfurt and Nürnberg developed a unique, especially for its time, style of painting with intricate renderings of insects and plants, as in a much later scientific illustration tradition. Besides being a revolutionary as a woman artist, her choice of subjects for her paintings were unconventional, insects generally being regarded as ugly (the prevailing belief was that they were spontaneously generated from mud) and not a subject for serious art. Even more striking was her rendering different stages of an

---

[13] Maria Sibylla Merian, 1647–1717, German. Artist and painter of natural history. She is known especially for her paintings of insects, first in her native land and later on a two-year trip to Surinam in South America. In addition to drawing and painting them, she also studied carefully insects, plants, frogs, snakes, and spiders.

insect's life, the caterpillar, pupa, and adult butterfly all on the same canvas. The entire life cycle of the creature was shown at once, to be grasped in an instant (Figure 7.5). The technology of time lapse photography of later centuries would make possible the complementary depiction of a single insect followed through all the stages from egg to its adult shape in chronological sequence.

Figure 7.5 A painting of caterpillar and butterfly by Maria Sybilla Merian (1705, illuminated copper engraving from a book in Senckenberg Naturmuseum, Frankfurt, Germany, Hannes Grobe). <http://upload.wikimedia.org/wikipedia/commons/thumb/c/c9/Merian-grafic-senkenberg_hg.jpg/728px-Merian-grafic-senkenberg_hg.jpg>.

Astronomy provides an analogous example. The time scale of our own lives being incompatible with that of stars, we study instead a whole sample of stars in our own lifetime, all of the same type but at different stages of their evolution, to provide the entire life history of a single star of that type. Such complementary descriptions using either a flow in time or a static, 'frozen', time-independent one, and the various discussions in this chapter all raise the question of how essential time is to understanding physics. Already in classical electricity and antenna theory, the complementarity between time and frequency was well recognized. Indeed, in a very definite way, these Fourier conjugates are related so that a distribution sharply peaked at an instant is spread broadly over the entire frequency spectrum, and vice versa. Quantum physics, through the entry of $\hbar$, the proportionality between energy and frequency, extended this to energy and time providing complementary pictures. Standing and travelling waves are both equally valid routes to understanding the same phenomena (Sec. 7.3.1).

Atomic spectroscopy provides a similar picture. The stationary states of an atom, solutions of the time-independent Schrödinger equation with the Hamiltonian H of the atom (Sec. 1.2.5), provide a complete picture of all possible phenomena associated with that atom. They are a complete set and, through superpositions of them, time variations, including scattering dynamics, can also be described. Indeed, for the first many decades of quantum physics and the study of atomic structure and dynamics, more and more precise spectroscopy with higher and higher energy resolution and accuracy was developed. It is only within the past decade that the advent of shorter and shorter laser pulses, now down to a few hundred attoseconds, has led to the alternative emphasis on direct integration of the time-dependent Schrödinger equation. Such a short pulse of course excites an atom into a superposition of a large number of energy eigenstates.

It comes down, therefore, to a question of alternative representations or points of view that has always characterized the subject of physics, especially quantum physics (Sec. 2.2). The same is true when it comes to time. Instead of arguing for the primacy of a time-independent (Julian Barbour) or time-dependent (Lee Smolin[14]) analysis, which are but limiting cases at two ends of a continuum, the picture and philosophy of

---

[14] Lee Smolin, 1955, American. Theoretical physicist, in quantum gravity and cosmology.

alternative representations and views says that both are equally capable of capturing the full or total world view, and neither is more or less. It is a matter of taste, and sometimes one of practicality, depending on the observer's location in space and time (as in whether a 17th-century artist working with paints on canvas one afternoon or a modern photographer with high-speed time-lapse techniques to follow the entire life cycle of a monarch butterfly), which is more convenient for the purpose at hand. Does a butterfly philosopher (Chapter 2's Chuang Tzu or contemporary Thomas Nagel[15]), with its constant flitting and swift flight in three dimensions, carry a faint trace of memory to wonder what it feels like to be a caterpillar, with its stillness or slow crawl in two dimensions?

Indeed, as in the discussion of frame transformations in Sec. 2.2.2, one can choose a judicious mix of representations to understand (riding alongside a photoelectron from an LS-coupling region at the small r of its birth to the jj-coupling at asymptotic distance when it is detected) the phenomena under investigation. Indeed, a useful mathematical technique of our times called 'wavelet analysis' uses as its basis finite intervals in both time and frequency rather than working in either of the Fourier conjugates of time alone or frequency alone. Musical notation long ago adopted such a hybrid picture. And, in the visual arts, cubism depicted simultaneously different profiles, front and rear views of an object, including a human figure, as a way of capturing the whole.

Physics deals with establishing correlations between phenomena and events, and can proceed without explicit invocation of time. The philosopher Soren Kierkegaard[16] made a profound observation: 'You can only understand life backwards but we must live it forwards'. This might be extended to say that physics itself can do away with time, but physicists, as time-bound creatures from birth to death, are condemned to view the physical world in terms of time. And, it might be added, in terms of classical concepts such as position, velocity, etc., which are not the ones that the underlying reality is constructed out of (see also Sec. 8.5).

---

[15] Thomas Nagel, 1937, American. Philosopher, known for his philosophy of the mind and his criticism of reductionist accounts. His essay 'What is it like to be a bat?' is widely known, as well as his recent criticisms of the neo-Darwinian view of natural selection as inadequate.

[16] Soren Aabye Kierkegaard, 1813–55, Danish. Philosopher, theologian and social critic, regarded as the premier existentialist philosopher.

8

Complexity and Emergence

8.1 Complexity

The world around us is varied and complex, perhaps even in reality, but certainly at least seemingly so. But our brains seem capable of understanding only in simple terms. Therefore, whether it is early man with mythical explanations or scientists giving natural explanations of that physical and biological world, a vast amount of knowledge and experience is condensed into simpler forms that we can comprehend. The whole quest of science is to take what looks complicated but reduce it to a few basic principles and inputs, to 'divine the rules of the game', much as one would watch several games of chess and see through all their complexity the basic underlying rules, moves, and strategies.

Thus, Newton explained much about motion in the Solar System through two principles, that of inertia and the $1/r^2$ force of gravity. These natural explanations may seem counter-intuitive, that instead of angels beating their wings incessantly along the orbit to keep the planet moving, the force acts perpendicularly to the realized motion at no energy cost (Figure 1.12). The myriad of slightly different snowflakes, seemingly the result of a God or Santa's army of elves stamping them out in a snowflake foundry, is seen instead as natural six-fold symmetry of aggregation in the crystal growth of water, with inevitable fluctuations making for small differences in the individual flakes (Figure 5.2). Both examples also show a fundamental feature that, in addition to basic principles, initial conditions or fluctuations not reducible to them are an ingredient of scientific explanation. They may be reducible to other principles in a later, more embracing theory, but there will then be new initial conditions. Seeking to eliminate them altogether in some Theory of Everything, or to take recourse to an 'anthropic principle' that makes our asking these questions intrinsic to the way things are, seems antithetical to science (Sec. 5.2.6). Even if we never quite reach it, and can expect only to get closer to it, there is an underlying reality independent of us. After all, there was a Universe and an Earth, with laws and principles governing them, long before the emergence of us or our ancestral hominids, mammals, or life itself.

Each area of physics has its small set of variables and parameters in terms of which observed phenomena are inter-related or explained in simple terms. Thus, in studying an ideal gas in a container, it is the volume, or the temperature when a finger or thermometer is stuck into it, and such so-called 'extensive' quantities that are of interest, not following the motion of each of the enormous number of molecules constituting the gas. Even if it were possible to follow all of them in detail, it would be irrelevant if one is interested in just one number, the temperature of the gas. After Maxwell and the kinetic theory of gases, the connection between the microscopic and macroscopic is available through the subject of statistical mechanics but thermodynamics is a self-contained subject in itself and in terms of its own concepts.

The same is true at the microscopic level. For studying atomic and molecular properties, covering an energy range from meV to a few tens of eV, the quark and gluon constituents of nucleons in the nuclei do not need to be considered. Indeed, it would be silly to do so even if it is true that all matter ultimately is a mix of quarks, gluons, and leptons, and their interacting quantum fields. This is also true of much of nuclear physics, where it is sufficient and appropriate to consider a nucleus as a collection of nucleons, protons, and neutrons, without invoking sub-constituents of them. Many of those degrees of freedom are frozen at the energies of interest so that they are irrelevant and thus ignorable.

And, even within atoms themselves, in today's field of cold collisions, a few parameters, the scattering lengths, are all that are necessary, whether it be lithium or rubidium that is under study. For decades, much effort has been devoted in precision spectroscopy or theoretical atomic structure calculations to understand each atom and differences between atoms, especially difficult the more the number of electrons that are involved. But none of these is relevant to studies at nanokelvin[1]

---

[1] William Thomson, Lord Kelvin, 1824–1907, British. Physicist and engineer, who made key contributions to electricity and to the understanding of heat energy and the Second Law of Thermodynamics. He established the absolute temperature scale used in the sciences, now named after him. He also contributed to telegraphy, especially in the

energies, a single parameter being sufficient to describe these alkali atoms.

In all this, an important consideration is the scale of energies involved and the corresponding inverse relation to lengths, an aspect of the uncertainty link between lengths and momenta. This gives rise to natural hierarchical levels of phenomena and explanations thereof, new collective coordinates and collective quantum numbers, and even new concepts becoming relevant at each level. This seems to characterize physics, and science as a whole. Identifying the few relevant variables at each level is essential for progress in physics.

8.2 Temperature

An especially nice illustration of such 'emergent' concepts in physics is temperature, T. It is necessarily a collective concept, not applicable to a single atom or molecule, but a measure of the random kinetic energies in a large collection of them. On the one hand, save for the Boltzmann[2] constant, k, which mediates the conversion from thermal energies to mechanical, there is no other input but the mechanical energies, $mv^2/2$, averaged over the random motions of the molecules in an assembly. But, on the other hand, any one of those molecules, upon isolation, cannot be said to have a temperature, and the concept itself does not even apply to it. A particle or a well-collimated beam of them, all with some definite velocity, may have very high kinetic energy but it would be inappropriate to say it has a temperature $mv^2/2k$. Only the random energies, and a collective average of them, is relevant to T. Indeed, there are current attempts through precision measurements to fix the value

laying of the first trans-Atlantic telegraph cable when he solved many of the problems as they arose, and to improving the nautical compass. He was the first UK scientist to be made a baron.

[2] Ludwig Boltzmann, 1844–1906, Austrian. Philosopher and theoretical physicist with major contributions to the kinetic theory of gases, thermodynamics, and statistical mechanics. Disorder and the concept of entropy and the counting of microscopic states for understanding the Second Law of Thermodynamics, so that it is a statistical law, are among his greatest contributions, as is his seeing the underlying atomistic structure of matter behind it. In this, he was opposed by much of the prevailing philosophy of physics around him that stressed energy and continuous distributions rather than a discrete particulate one. The Boltzmann equation and the fundamental constant, k (introduced actually by Planck), relating energy and temperature carry his name. He is also seen as a pioneer in understanding free energy, a concept of great import to biology.

of k and thus reduce T to an energy, much as fixing the speed of light, c, renders measurements of length to those of time.

As with other instances of concepts that are relevant at one hierarchy not being simply reducible to those at a lower level, there is often a suppression of a large number of degrees of freedom in this reduction. While present, they are irrelevant, and are frozen out. At the level of atomic interactions in the energy range of a few eV or less, quark degrees of freedom are simply not excited, being at a much higher energy scale and thus ignorable. Similarly, while there is an Avogadro[3] number of degrees of freedom of each molecule, only the one of temperature is relevant in the thermal averaging, which makes sense of course only under the so-called ergodicity that pertains to thermodynamic equilibrium.

8.3 Phases and Phase Transitions

Everyone is familiar with phases of matter. Thus, water exists either as gaseous water vapour, or as liquid water, or as solid water-ice (actually ice itself exists in many phases, one, Ice IX, made famous in a literary work! [33]). At the level of the individual molecule, H_2O, there is no distinction and no phase information resides in it. But, in the aggregate, and depending on external conditions of temperature and pressure, entirely different phases emerge. And, at certain transition values of these external parameters, different phases can co-exist in equilibrium. Again, it is a fact familiar to everyone that at normal atmospheric pressure, water and ice co-exist at $0°C$, and water and steam at $100°C$, these even defining the temperature scale. Less familiar to the layman, but well recognized in physics, is a 'triple point', when all three phases co-exist at a specific temperature and pressure! (This triple point of water at $0.01°C$ or 273.16 K is used to fix the temperature scale.) Such phases and phase transitions between them are another good illustration within physics itself of emergent concepts and phenomena.

---

[3] Lorenzo Romano Amedeo Avogadro, 1776–1856, Italian. Mathematician and scientist who contributed to understanding the molecular structure of gases, and to clearly distinguishing between atoms and molecules. The number of molecules in a 'mole' of any gas, defined as the mass in grams equal to the molecular weight, a large number, approximately 6×10^{23}, is named after him.

8.4 Even More Profound Emergences

Temperature might serve as a good model for other emergent concepts that play a role in our lives. The meaning of a text is not in any of the words constituting it and certainly not in the letters, all texts being made of the same letters of the alphabet and a few other symbols. Our scientific papers today and a book such as this are often set in lines of some TeX compiler, the letters of the alphabet and a few other strokes capable of rendering astonishingly varied fonts and symbols. In today's digital displays, arrangements of seven line strokes can reproduce any letter or number. But the text itself and its meaning are not to be reduced to such.

Or, as another example, the use of coins, currency, and other financial paper, now disembodied in electronic space (see Sec. 4.3), has as its essence a transaction between two parties for goods and services. When you buy something and pay at the store counter, say with a cheque, when and how are you actually paying for it? Is it at the moment you sign the cheque and hand it over to the store clerk? Is it when, at the end of the day, the store bundles up and presents the cheques to a local bank branch? Is it when, through a series of intervening banks and financial institutions in between, the funds are deducted from your own account, where, perhaps, an earlier direct deposit from your employer for services you rendered over the past week or month credited you with your salary? We recognize that much of this is not essential, although diffusively spread out and not easily traced or accounted for, for the essence of the transaction itself, the trade in goods or services.

Life and consciousness, whether or not reducible to certain neural circuits and cells in some yet-to-develop understanding (and, surely, only in some distant future, notwithstanding the enthusiasm of some current-day neuroscientists!) may well be like temperature and phase in that even when there is nothing more to be invoked, nevertheless they are emergent concepts not present in or even relevant for those individual constituents. Temperature may be an example of physics lending an appropriate metaphor to these other disciplines. Populations are collections of individuals, and an individual organism a collection of cells, molecules, and genes, but whether in sociology or biology, properties and characteristics of the aggregate, while not involving any new laws of science, are nevertheless not simply contained at these smallest levels.

This may be the important lesson that the admittedly simpler examples of physics, such as temperature and phase, have to offer to the more complex emergences of biology or sociology.

The fundamental rules themselves may even be trivial, and it is in their collective realization that richness may lie, just as in the game of chess. The game of 'Life' provides another nice example. Invented by the mathematician John Conway[4], a few simple rules for cells on a two-dimensional grid are the axioms of the game. They govern whether a cell is occupied or not for the next iteration, given the configuration at the previous iteration. Yet, starting from different initial patterns, an astonishing range of patterns, some even dynamic with a pattern sliding across the grid, are generated [34]. Some of these had not been anticipated even by the 'creator' of the game, and new patterns continue to be investigated. Similarly, my painter friend who does large canvases of abstract compositions of colours said that he was surprised when someone saw an alligator in one; once pointed out, he himself could see 'a full alligator, from tail to snout' [35].

Biological life and the distinction between a living and a dead organism are orders of magnitude more complicated than the examples from physics. It is usually easy to distinguish a dead body from a living animal, sometimes just moments before and after a specific moment of transition, all the physical elements unchanged. That moment of transition may not be marked by any change in the physics but clearly is profound. For long in science's history, organic and inorganic chemistry were viewed as qualitatively distinct, with a *vis viva* or life force essential in the former. But that barrier was finally broken when it was shown that organic compounds can be synthesized from the inorganic components of carbon, hydrogen, oxygen, etc. without any other input invoked. This provides a lesson that, similarly, those who favour a sharp distinction for life, including the invocation of something beyond science in some religious belief, may be prematurely giving up on an explanation within science itself.

On the other hand, scientists need to display the humility required by the fact that current science is certainly far from identifying what

---

[4] John Horton Conway, 1937, British and American. Mathematician who has contributed to finite groups, knot theory, game theory, and number theory. Known also for his invention of 'surreal numbers' and an arrow notation for handling extremely large numbers, and for the game of 'Life.'

characterizes life, the temperature or phase analogies being baby steps in comparison. But there is no need to preclude such a search. Indeed, for both sides, it is but hubris to think that, in our own times, will be found a Theory of Everything. Both such a claim by scientists, or by philosophers or religions to say some supernatural, extra-scientific input is necessary, are unjustified if not untenable. Just as Copernicus[5] dethroned our Earth from any special place in space, so also in terms of time our own 100-year life span must be recognized as nothing special. The quest for scientific understanding is ongoing and endless; 'the end of science' is but an oxymoron. So long as scientists exist to ask questions, science will continue. The essence of the Copernican principle of there being nothing special about us or about our lives and times should give an acceptance that in some distant future we will understand more the emergence of life and consciousness from physics, chemistry, and biology. But for today, some of the ideas behind emergent concepts such as temperature and phase are pointers to even more sophisticated understanding of emergences.

8.5 Classical Physics Itself an Emergence

What does physics deal with? The subject started with Galileo and Newton as describing the motion of material bodies, thought of as point masses, particle mechanics. Knowing the state of the particle at some instant meant knowing its position and velocity. The laws of physics (motion) would then describe the subsequent time evolution by providing position and velocity at later instants once the forces were specified. But, even already in that, physics was clearly not restricted to just that particle but applied equally to any other particle subjected to the same or similar conditions. Science itself would have little meaning without such universality in its application. Indeed, the very repeatability of experiments is an essential ingredient of any science, especially an experimental one! And the requirement that the same experiment done at a different place, translated from the original, or at a different time, must lead to the same physics, again essential to have any meaning or

[5] Nicolaus Copernicus, 1473–1543, Polish. Astronomer, renowned for his heliocentric system of planets orbiting the Sun, and for the larger philosophical idea, built on his dethroning of our planet's central position, that there is no privileged position in the Universe.

validity to the subject, leads to the most profound laws of conservation of energy and momentum (Sec. 5.1.2).

Later, the study of particle masses was supplemented by that of waves, whether mechanical or electromagnetic, also carriers of energy and momentum. Again, a wave equation describes the evolution of not just one particular wave but any other identical one under identical conditions. But it is true in classical physics, whether of particles or waves, that each can be tagged as with runners along race tracks and thus the motion of an individual particle or wave kept track of or predicted.

Next, it is also easily recognized that among many characteristics, some are of no significance to the physics. Thus, in describing the motion of a ball from the instant at which it is hit by a bat to the later time when it is caught in a fielder's hand, what physics accounts for is the connection between the two events. The specific positions attributed to the two will vary from one observer to another in the stadium. Those sitting on one side may describe the motion as a parabolic arc from left to right, the spectators on the other side as just the opposite, from right to left. There is no significance in this for physics. With the advent of Special Relativity, we also recognized that different inertial observers viewing the ball's motion may even ascribe different spatial separations and time intervals between the two events, only the space–time interval being an invariant among all such observers (Sec. 7.2).

Every viewer in the stands, or on blimps moving with uniform velocities over the stadium, will also ascribe some trajectory to the ball's motion. These will be parabolas but of varying tightness, with the limiting case of straight up and down motion as seen by an inertial frame moving with the same horizontal velocity as the ball. Therefore, the actual shape of the trajectory is not an element that physics has to or does explain, and we recognize this readily. We would see any 'Many Trajectories' interpretation of the motion of the ball, any claim that the ball actually executes all these multitude of trajectories as a somewhat strange and extravagant rendering of what is actually the ball's motion as observed by different observers. Every observer describes the motion from beginning to end in terms of a trajectory but the trajectories may all be different. There is no trajectory by itself; it is not a concept solely of the ball but of ball plus observer. The question of what the ball 'really' did has no meaning without asking 'as seen by whom?' Indeed, in the frame of reference of the ball, it is motionless and it is all the observers in the stadium and on blimps that execute a multitude of parabolic trajectories.

There is here an underlying reality of motion, but that motion is of ball relative to an observer and it is no surprise that there are different descriptions by different observers (among them, the ball itself). And the underlying reality of the entire motion, from bat to fielder's hand, does not rest on a time-dependent Newtonian description or a time-independent one such as Hamilton's in terms of an action integral, although it is equally amenable to either description. There is no time sequence or path, all possible ones (Figure 7.1), among which is the one that makes the action integral stationary, being distinguished.

The further step into quantum mechanics denies any meaning to trajectories or paths, concepts that require specification of both position and velocity but whose simultaneous specification is ruled out by the uncertainty principle. Even more, the concepts of position and momentum themselves do not apply, these too being valid only in 'the classical limit'. Even without going into quantum field theory, within quantum mechanics itself of even just a single particle, the state of a system is a complex valued wave function or ket. All other observables for it, position, momentum, etc., are derived from the wave function through certain expectation values. The notion of a wave or particle is also relevant only in the classical limit, as discussed in Sec. 2.3.

Further, mere labels have no meaning, a physical system being characterized only by physically measurable invariants, a ket labelled by the quantum numbers of the operators that commute with the Hamiltonian. A free electron is an object of a certain mass, charge, spin, energy, and linear momentum as defined by our classical measuring apparatuses capable of making or detecting it. Identical particles cannot be differentiated by merely labelling them with numbers on their backs as with runners or particles or waves in classical physics (Sec. 2.2.1), where race tracks or paths and passage through them, even when intersecting, have meaning. Indeed, under interchange of any pair of otherwise identical quantum entities, that is, interchange of those labels, the wave function must satisfy the Pauli principle of being either symmetric or antisymmetric, depending on whether the entities are bosons or fermions, respectively (Sec. 7.3.3).

With the physical system described by a wave function or ket, the Schrödinger equation, the counterpart of Newton's, gives a complete deterministic evolution of that state once the potentials are specified. It describes not any single particle or situation but a whole ensemble of them. Thus, for any single radioactive nucleus such as of radium, Ra, its decay with emission, say, of an alpha particle or an electron (beta decay) is entirely unpredictable. Starting with identical samples of such Ra nuclei, one may go off the very next instant, whereas another may not decay over the entire length of the Universe's lifetime. What is meant by the half-life of Ra is the time within which there is a 50% chance of seeing it decay. Or, in a large collection of such Ra nuclei, in that time roughly half will have decayed. There will always be some spread around that value, fluctuations around that average.

Similarly, turning to a motion in space, if one has a source at S, say of electrons, perhaps created by a radioactive decay, and a screen covered with detectors some distance away, and an intervening slit, then each observation when the source emits and at a later instant a detector receives is entirely unpredictable in terms of which detector on the screen fires. Because of how electrons are registered as a lump of charge, e, for a source emitting only one at a time, it is only one detector that fires, never more simultaneously, the electron being received at some one point on the screen. An electron, defined as a lump of charge, mass, and spin, is never detected simultaneously at two detectors, only at one at a time. But it could be any one of the detectors anywhere on that infinite screen. What the wave function of this one-slit arrangement provides is the probability for receipt of the electron on the screen, and it will be peaked at the spot on the screen that is in line with the one connecting the source to the slit, that is, the path that a classical description would have predicted as the unique one. The probability drops off away from that central spot but is non-zero everywhere, tailing off to zero only at $\pm\infty$. Again, if a large number of such observations is compiled, the intensity on the screen, where the charge is detected (but, again, the charge is never smeared out), will appear as in Figure 8.1.

For a different arrangement, say with two slits instead of one, again each individual electron from the source will be detected unpredictably by some detector on the screen as an electron. The wave function for this arrangement of source–double slit–screen will be different from the previous one. Again, it will determine the probability for receipt on the screen in any individual experiment or the total intensity for a large collection of electrons, emitted one by one. This pattern in Figure 8.2 is different from the one in Figure 8.1. One striking difference is that it is not just the sum of two intensity patterns for each slit with two peaks

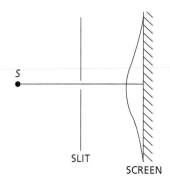

Figure 8.1 Schematic arrangement for a single-slit diffraction pattern, with a source, S, of photons or electrons, and a screen, with an intervening slit.

at two different spots corresponding to the path from source to each slit when extended to the screen. Instead, the peak intensity is at the middle of those two positions.

Another striking difference is that there are now spots on the screen with zero probability or intensity, the detectors at such spots never firing. The contribution to the wave function from the different slits mutually interfere destructively so that there is a zero of the wave function at these spots. These 'nodes' in the pattern and the patterns themselves in Figures 8.1 and 8.2 were familiar already in optics from Newton's time but they manifest equally for electrons or protons or even heavier carbon fullerenes ('buckyballs'[6]), as has been experimentally demonstrated. Experiments have also verified the random appearance of each electron on the screen but, together, that many such repetitions construct the diffraction or double-slit pattern (Figure 8.2). And there is no reason to doubt that this would also work with Mack trucks, that this is of universal applicability, reflecting the wave nature of all things, even those which have a particle as their classical limit.

The question of paths or the nature of detection falls into the following picture. With a strict focus only on what is actually measured, it is not meaningful to ask about things that are not. The electron is

---

[6] Named after Buckminster Fuller, 1895–1983, American. Architect and inventor, and 'futurist'. He is known for his building and popularizing the geodesic dome. A new form of carbon with 60 carbon atoms in a three-dimensional structure like a soccer ball was named 'buckyball' and, together with other carbon structures, classed as a group under the name 'fullerenes'.

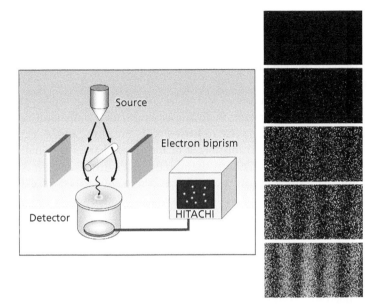

Figure 8.2 Double-slit arrangement. The set-up is otherwise as in Figure 8.1 but with two slits between source and screen. The left panel shows the experimental arrangement of electron source and detector with intervening bi-prisms that act as slits. The right panel shows the gradual build-up of the two-slit diffraction pattern from individual detection of the electrons. Courtesy of the Central Research Laboratory, Hitachi, Ltd., Japan, from the work of A. Tonomura *et al*, *Am. J. Phys.* **57**, 117 (1989).

observed either at the source or at a detector some time later, for any particular experimental set-up or apparatus. There is no meaning to unobserved entities such as the path followed in between. If indeed detectors were set up intermediately for this purpose, that would be a different set-up with a different wave function and a different pattern. Words such as 'paths' or 'trajectories', or questions about which slit the electron went through are just that, words, but with no place in physics.

The wave function for a particular arrangement of apparatus describes all the physics. Had we wave-function-measuring apparatus, there would be no ambiguity in keeping track of what it is at any location and at any instant (every location and every instant!). Quantum physics would give a complete description of it. But instead, we observe, measure, or even describe in terms that are classical. It may be counter to our classical intuition that an electron went through both slits or that having two slits open makes for some nodal points with the electron never seen at them while not so for a single slit. But all these words and ways of speaking are not relevant to the physics of one set-up or the other.

The wave function can be said to be a sum over all possible paths joining source to detector, including some (many!) outlandish ones that roam over the Universe in between. As noted in Sec. 7.2, Feynman's path-integral formulation of quantum mechanics does indeed construct the wave function as the sum over all paths. Each path contributes with some weight to the wave function and to the probability of receipt on the screen, while some of the outlandish ones or paths that hit the barrier on either side of a slit contribute negligibly or not at all. Paths close to the one we expect based on classical geometrical intuition contribute more, especially for objects that behave essentially classically, thus truer for Mack trucks than for electrons. Finally, that particular path in Figure 8.1 of a straight line from source to slit and extended to screen emerges as 'the' classical limit, Newtonian physics emerging as the limiting case of the quantum description. Indeed, already in the Lagrangian and Hamiltonian formulation within classical physics and before the advent of quantum physics, Hamilton's variational principle had so replaced globally in the action integral the local Newtonian view of step-by-step motion along that line, while being compatible with it, as discussed in Sec. 7.2.

The classical picture, whether for the entire motion or for an electron at source and at detector, is itself an emergent one, as is whether an entity is a wave or a particle. These are all concepts that are not part of the underlying reality of a quantum world. Just as we had already grown to accept in classical physics that concepts such as paths and trajectories do not have intrinsic meaning in Galilean or Einsteinian Relativity, but are dependent on the specifics of an observer or of a description, and that we need to be alert to this and not identify them with reality itself, now we have to extend that to even the concepts of position, momentum, etc.

Or to whether we have a wave or a particle, these being meaningful only in the classical limit (Sec. 2.3). The underlying reality is not in terms of them but in complex wave functions and states, and of wavicles (Sec. 2.3). But, as classical beings, and our concepts and words themselves having evolved from those classical experiences and intuitions, we cannot escape having to use them in our thinking and communication, just as everyone in the stadium or a blimp ascribes a trajectory to the ball in talking about it. But we have to be alert to the fact that underlying reality is not written in that language. It is, therefore, like being condemned to read a text in a language no longer directly accessible but now only in translation.

Alternatively, it is as if we always view the world with distorting spectacles. Interestingly, Newton himself, as the developer of geometrical optics, knew that our eyes image the world as inverted on the retina. It is our neural processing that converts this inverted two-dimensional image of our twin eyes to the upright three-dimensional appearance of the objects around us. Much of the discomfort or difficulty in grasping quantum physics and some of what are posed as paradoxes lies in not keeping clear these matters and extrapolating from our models to reality itself.

What, then, is involved in the emergence of the classical world from an underlying, inherently quantum reality? As in other examples of emergence, certainly there is a tracing over a large number of degrees of freedom. The quantum wave function, even of a single point particle, is a function of a complex variable. A quantum spin-1/2, as we saw in Chapter 4, already has a three-variable space of an enormous number of states, even though observations see it as just one of two things, up or down. With more particles, the dimension of the space involved explodes and in any interaction with an observing apparatus, most of them are traced over.

In particular, phases of complex functions vary continuously from 0 to 2π, and are very delicate (Sec. 4.2.1), subject to disturbances from interactions with the external world, itself of an enormous number of degrees of freedom. So, scrambling of these phases is certainly one element of the emergence, as is the fact that, generally speaking, getting to the classical limit involves larger aggregates of particles, although special cases such as superconductivity can retain meaning in a macroscopic phase. And, even for a single electron, when it is realised as a bundle of a certain amount of electric charge, mass, and spin angular momentum when emitted or absorbed by an apparatus, coherences of its wave function have been scrambled for it to so appear as a particle.

Again, using an earlier example as an analogy, temperature emerges as a single number upon averaging over the random kinetic energies of an Avogadro number of particles. Although no new element of physics needs to be invoked in the passage from the statistical mechanics description of those particles' motion to the thermodynamic limit's T, the passage back is, of course, impossible. In a way, questions such as the emergence of classical wave and particle, position, or other observable from the underlying quantum world and using them to understand that world is like having only thermometers through which to grasp the individual molecules' motions. The quantum wave function contains an enormously larger amount of information than what is realized in our observations with our slits or other apparatus.

The essence lies in what is meant by the state of a physical system. In quantum physics, it is described by a complex wave function or, as in Sec. 2.3, a Dirac ket, the ket labelled by the values of the quantities that commute with the Hamiltonian. No other labels have meaning, not mere number tags that correspond to nothing in physics nor a position in space, x usually not commuting with H. For an electron in its pattern on a screen of a slit assembly, single or double, or for it in a particular state of the hydrogen atom, no single position applies. When its location is sought, it may be found, but that may be at any point on the screen or in space (except exactly at one of the nodes).

If an apparatus is churning out identical copies of hydrogen, say from a chemical reaction at the molecular level resulting in atomic hydrogen in the ground state, definite values can be ascribed only to measurements of energy and angular momentum, that each copy will have only -13.6 eV and zero, respectively. But, if it is position that is measured, the electron in each copy may be found at any location randomly, and only a probability distribution can be prescribed for it. On the other hand, one can talk of an electron in any of these systems as being at a precise location, for instance when it is located at the tiny detector, in principle arbitrarily small (within non-relativistic quantum mechanics), that fired, but then it is not in a definite state of energy and angular momentum but in a very large superposition of such states. An electron as a particle at its emergence from a source or at a detector, defined as a lump of charge, mass, and spin angular momentum, is an object in the classical limit that has averaged over a large number of quantum states.

The step into quantum field theory, currently the closest model to the underlying reality, compounds but also illuminates these matters further. There are interacting quantum fields and it is excitations in them that we observe as electrons or other entities (Sec. 7.3.3). Thus, in a radioactive decay of even a single nucleus or neutron, we have the fields of electron, neutrino, and nucleon (or quark) to consider. The initial state is of one excitation from the vacuum of the nucleon field, say, as a neutron. The final state is a combined one of one excitation of that nucleon field in the form of a proton, one excitation of the electron field, and one of the anti-neutrino. There is no specific location or instant (of decay) for going from initial to final state. The fields themselves exist over all space, space and time being only grid parameters over which the field functions are defined. Each individual observation of a single-slit set-up corresponds to an electron appearing at a location (the source's exit slit) at some unpredictable instant and an electron being absorbed at a detector, again which one unpredictable, possibly by an inverse beta process again involving the three fields.

Quantum physics describes this and it embraces not a single event but all the myriad of them when an electron leaves the source and a detector on the screen receives it. With each repetition of the experiment, sometimes one, sometimes another detector fires. It is from an underlying quantum field (or interacting fields) that, first, a non-relativistic wave function emerges as the result of field operators acting on the vacuum state and, next, the electron is seen as a particle in the classical limit that scrambles phases and coherences. This happens at source and detector for each repetition of the experiment. Each instance is unpredictable, the statistical nature as in the build-up of a pattern in Figure 8.2 residing in the wave function (and probability interpretation: Sec. 1.2.2) that encompasses all possibilities but only one being realized in each repetition.

The wave function of quantum physics does not describe a single experiment but the two-slit pattern, just as it is not the decay of a single radioactive nucleus but beta decay as a process that is described. To the charge that the theory then is incomplete in not applying to individual elements, and that a future theory will also provide that, the problem is that all such attempts at extensions, which go under the name of hidden-variable theories, lead to predictions in conflict with experiment. Hidden variables in each nucleus whose knowledge would allow us to say the precise time at which each decays, or an underlying point particle that is guided by the wave function across slits, lead to predictions in conflict with what we know of radioactive decay or interference patterns on a screen. To date, all experiments and observations suggest that there is no such extension, that quantum physics is indeed today's complete model of the underlying reality. Rather, we must accept that what we refer to as an individual experiment or a particular nucleus decaying is meaningful only as a classical limiting case, averaged over the underlying quantum reality. The very feature of merely tagging an individual with a number label slapped on the back as on runners in a race, nucleus 1, nucleus 2, etc., and attributing significance to it, is not an element of that quantum world.

Sources and sinks (detectors) are large aggregates and while themselves part of the quantum world, have a large scrambling of the phases involved of the emitted or absorbed entity, and it is here that the electron is manifest as a classical entity of some mass, charge, etc., not at points or times in between. Identical particles, when more than one are present, are, in a field theoretic picture, simultaneous double or triple excitations of the field, and Pauli principle requirements (Sec. 7.3.3) are natural given the intrinsic properties of the operators involved in such excitations out of the vacuum. There are no exact locations in space or time for the electrons created and the non-locality of the subject of quantum physics is inherent to it. As noted in Sec. 6.3, the rendering with just a few parameters of an underlying element that itself resides in a huge-dimensional space necessarily leads to non-locality.

The picture we have today is that deep, deep down, the underlying 'real' world is closest to our current understanding as a mess of interacting relativistic quantum fields. (Even that is not reality itself, but our current best model.) But, we observe and experience that world at various levels and hierarchies, and do so in terms of emergent concepts and measures that are appropriate for those purposes and levels. It would be foolish to do otherwise. Field theories may be appropriate at times, non-relativistic quantum mechanics at others, and the older, non-relativistic classical mechanics may be, and is, the appropriate one for many purposes. The very concepts of position, momentum, space, and time are equally derived or emergent, much of that because of our own existence as lumbering, macroscopic objects shaped by the physical, chemical, and biological evolution of the Universe that has led to us. And, of course, we can tag runners on their backs as a meaningful classical feature!

Time and time again, the history of physics has sounded cautionary notes on our possibly being misled by those various elements that have shaped us. For centuries, it had seemed that being at rest was a natural state to which all objects in motion around us tended, but this was only because of the unseen frictional forces ubiquitously around us. It was only after Galileo and Newton that it became clear that the proper association was between forces and accelerations, not velocities. And then for another long period, their concepts of space and time seemed obvious till the Special Theory of Relativity disabused us of that. The slicing of a combined space–time into space and time is itself dependent on the inertial frame or observer, and all these frames are equivalent in physics. The very concepts of spatial separations and time intervals are not universals, however intuitively so they may seem (as they had in all the experience with non-relativistic speeds), but depend on the inertial frame of the observer.

Quantum physics went further in removing the very primitives of our constructs such as position and momentum, wave and particle, from being the ones of the underlying reality. But, on the other hand, the feature of emergence of concepts and measures at each level of hierarchy show also how they continue to be relevant for descriptions at that level. Therefore, even today, for most motion, classical Newtonian non-relativistic physics continues to apply, whether in everyday lay experience or even for the rocket engineer. With v/c small, inertial observers can agree on the time interval as a Newtonian absolute common to all. And, even in relativistic quantum contexts in laboratory settings, we are compelled to use the only language and concepts we have, such as position, momentum, etc. There is not only no harm in this but we must recognize it as essential at that level of hierarchy, while also not mistaking the model itself for the reality.

All of this is a part of the tension that is intrinsic to our subject, revolutionary upheavals co-existing with a conservative persistence of the old. Every extension to new regimes far from everyday experience, into the very small and very energetic, has opened a new perspective that also fundamentally changed our understanding of the primitives of our subject. These changes in the fundamentals apply, of course, to everything, including the large and unmoving, immovable mountains. They, too, are made of atoms, with electrons inside moving at relativistic speeds under quantum principles. Yet, effective theories and concepts that emerge at successive hierarchical levels account for the continuing validity of much that had lasted for centuries without that recognition of deeper levels. This had to be so, given the extensive interlocking pieces of evidence that had built the scientific edifice over those centuries. Together, a coherence of the subject has, and continues to be, maintained across the past five centuries. Different representations of an underlying reality are the essence of physics and of our understanding of the world around us through them.

References

1 Charles Darwin, *The Variation of Animals and Plants Under Domestication* (John Murray, London, 1868), Sec. 1.6.

2 Steven Weinberg, *The First Three Minutes: A Modern View of the Origin of the Universe* (Basic Books, New York, 1977).

3 R. P. Feynman, *QED: The Strange Theory of Light and Matter* (Princeton University Press, Princeton, 1985).

4 R. E. Peierls, *Surprises in Theoretical Physics* (Princeton University Press, Princeton, 1979).

5 D'Arcy Thompson, *On Growth and Form* (Cambridge University Press, Cambridge, 1943).

6 Steven Weinberg, *Gravitation and Cosmology: Principles and Applications of the General Theory of Relativity* (John Wiley, New York, 1972).

7 Brian Greene, *The Elegant Universe: Superstrings, Hidden Dimensions, and the Quest for the Ultimate Theory* (W.W. Norton and Company, New York, 2003).

8 Paul J. Nahin, *When Least Is Best: How Mathematicians Discovered Many Clever Ways to Make Things as Small (or as Large) as Possible* (Princeton University Press, Princeton, 2004).

9 R. P. Feynman, R. B. Leighton, and M. Sands, *The Feynman Lectures on Physics* (Addison-Wesley Press, Reading, 1965), Volume 1, Chapter 19.

10 E. Gerjuoy, A. R. P. Rau, and L. Spruch, 'A unified formulation of the construction of variational principles', Rev. Mod. Phys. **55**, 725–774 (1983).

11 V. Weisskopf, 'Of atoms, mountains, and stars: A study in qualitative physics', Science **187**, 605–612 (1975).

12 P. Kustaanheimo and E. Stiefel, 'Perturbation theory of Kepler motion based on spinor regularization', J. Reine Angew. Math. **218**, 204–219 (1965).

13 F. Iachello and A. Arima, *The Interacting Boson Model* (Cambridge University Press, Cambridge, 1987).

14 B. R. Judd and G. M. S. Lister, 'Selection rules in the atomic f shell from quarklike substructures', Phys. Rev. Lett. **67**, 1720–1722 (1991).

15 J. C. Taylor, *Gauge Theories of Weak Interactions* (Cambridge University Press, Cambridge, 1976), Chapter 13.

16 A. R. P. Rau, 'Extra dimensions to remove singularities and determine fundamental constants', Am. J. Phys. **53**, 1183–1186 (1985).

17 Burton Watson, *The Complete Works of Chuang Tzu* (Columbia University Press, New York, 1968).

18 U. Fano and A. R. P. Rau, *Atomic Collisions and Spectra* (Academic Press, Orlando, 1986).

19 A. R. P. Rau, 'The asymmetric rotor as a model for localization', Rev. Mod. Phys. **64**, 623–632 (1992).

20 D. G. Truhlar, A. D. Isaacson, and B. C. Garrett, *Theory of Chemical Reaction Dynamics*, Vol. 4, edited by M. Baer (CRC Press, Boca Raton), p. 65.

21 M. A. Nielsen and I. L. Chuang, *Quantum Computation and Quantum Information* (Cambridge University Press, Cambridge, 2000).

22 D. Uskov and A. R. P. Rau, 'Geometric phases and Bloch-sphere constructions for SU(N) groups with a complete description of the SU(4) group', Phys. Rev. A **78**, 022331 (2008). A. R. P. Rau, 'Mapping two-bit operators onto projective geometries', Phys. Rev. A **79**, 042323 (2009).

23 R. Aaij *et al.* (LHCb Collaboration), 'First observation of CP violation in the decay of B_s^0 mesons', Phys. Rev. Lett. **110**, 221601 (2013).

24 A. R. P. Rau, 'R. A. Fisher, design theory, and the Indian connection', J. Biosci. **34**, 353–363 (2009).

25 T. Beth, D. Jungnickel, and H. Lenz, *Design Theory* (Bibl. Institut, Zürich, 1985).

26 J. C. Baez, Bull. New Ser. Am. Math. Soc. **39**, 145 (2001), and www.jmath.usr.edu/home/baez/octonions

27 A. R. P. Rau, 'Supersymmetry in quantum mechanics: an extended view', J. Phys. A: Math. Gen. **37**, 10421–10427 (2004).

28 R. W. Haymaker and A. R. P. Rau, 'Supersymmetry in quantum mechanics', Am. J. Phys. **54**, 928–936 (1986).

29 Tom Koch, *Cartographies of Disease* (ESRI Press, Redlands, 2005).

30 Denis Wood and John Fels, *The Natures of Maps* (University of Chicago Press, Chicago, 2008).

31 A. R. P. Rau, 'Cross between Born and WKB approximations: Variational solutions of nonlinear forms of the Schrödinger equation', J. Math. Phys. **17**, 1338–1344 (1976).

32 Julian Barbour, *The End of Time: The Next Revolution in Our Understanding of the Universe* (Oxford University Press, Oxford, 1999).

33 Kurt Vonnegut, *Cat's Cradle* (Holt, Reinhart and Winston, New York, 1963).

34 Martin Gardner, 'Mathematical games', in *Scientific American* (October 1970).

35 Randell Henry, painter, Southern University, Baton Rouge, personal communication.

Index